WITHDRAWN

# THE SOLAR HOUSE

**BOOKS FOR LIVING WISELY FROM *MOTHER EARTH NEWS***

*Mother Earth News* came into being when astronauts were flying to the moon and young people were questioning the status quo. People began to seek answers in Mother Nature. They turned back to the land, seeking a simpler, more grounded, and more fulfilling way of life.

Times change, and so do fashions and phrases. The world is still full of questions, but the answers are more likely to involve concepts of ecology, interconnectedness, and sustainability. No one talks of going "back to the land" much, but we still find answers in Nature, and in our own natures.

The partnership between *Mother Earth News* and Chelsea Green has a common taproot in the Earth. The Books for Living Wisely series will reflect the common mission of the two companies: to introduce new people to ideas that are at once exciting, important, and basic.

Bryan Welch                              Stephen Morris

Publisher, *Mother Earth News*           Publisher, Chelsea Green

# THE **SOLAR** HOUSE

## PASSIVE HEATING
## AND COOLING

Daniel D. Chiras, Ph.D.

CHELSEA GREEN PUBLISHING COMPANY
WHITE RIVER JUNCTION, VERMONT

Printed in Canada.
First printing, October 2002.

05 04 03 02   1 2 3 4 5

Printed on acid-free, recycled paper.

Due to the variability of local conditions; materials; skills; site; the complexity of building a home in general and a passively conditioned structure specifically; and so forth, Chelsea Green Publishing Company and the author assume no responsibility for personal injury, property damage, or loss from actions inspired by information in this book. Always consult manufacturers of specific products, applicable local and state laws, and the National Electric Code. When in doubt, ask for advice. Opinions expressed in this book are based on the author's personal experience.

Many of the designations used by manufacturers and sellers to distinguish their products are claimed as trademarks. Where those designations appear in this book and Chelsea Green was aware of a trademark claim, the designations have been printed in initial capital letters.

**Library of Congress Cataloging-in-Publication Data**
Chiras, Daniel D.
    The Solar House: passive solar heating and cooling / Daniel D. Chiras.
        p.      cm.
        Includes bibliographical references and index.
        ISBN 1-931498-12-1
    1. Solar houses. I. Title.

    TH7414.C475 2002 2002031482
    697'.78—dc21

Chelsea Green Publishing Company
Post Office Box 428
White River Junction, VT 05001
(800) 639-4099
www.chelseagreen.com

**DEDICATION**

*To Linda, my love*

# TABLE OF CONTENTS

# ACKNOWLEDGMENTS

A GREAT MANY PEOPLE have aided me in the preparation of this book. I am deeply indebted to those who answered my frequent questions and provided additional information to help create a comprehensive and accurate book on passive heating and cooling, including Ron Judkoff and Chuck Kutscher (NREL), Doug Hargrave (SBIC), Alex Wilson (BuildingGreen), Bill Eckert (Friendly Fire), Niko Horster (Chelsea Green), Bruce Brownell (Adirondak Alternate Energy), Heinz Flurer (Biofire), Vashek Berka (Bohemia International), Randy Udall (CORE), James Plagmann (HumanNature), and David Adamson (EcoBuild). Many thanks to the dozens of individuals and companies who provided photos for use in this book.

I am especially thankful to those who read early drafts of the manuscript, offering insightful comments and vital corrections: Niko Horster, Jim Schley, Ron Judkoff, and Bill Eckert. Your assistance was invaluable! I also owe a debt of gratitude to the folks at Chelsea Green, including my editors Jim Schley and Alan Berolzheimer for helping me shape and refine this book and to my illustrator, David Smith, for his fine contributions to this project.

Finally, on a personal level, many thanks to my two darling sons, Forrest and Skyler, for being themselves—insightful, funny, joyful, hard-working, and responsible—and to my partner, Linda, for her patience, understanding, kindness, and unwavering love and support.

# INTRODUCTION: ACHIEVING COMFORT IN ANY CLIMATE

**A FEW YEARS AGO,** a friend of mine called and asked if I would give my opinion on a home she was considering buying. The house in question was a passive solar design—that is, a home that is heated by the sun directly through south-facing windows. I was living in a passive solar home at the time (see figure I-1) and had been studying such designs for twenty years, so I agreed to go along to assess the structure.

The next morning we headed out to take a look at the house, located down a twisty gravel road about half an hour from my home in the foothills of the Rockies. It was a picture-perfect fall day, a great time for evaluating a solar home.

**FIGURE I-1**
*The author's first passive solar house offered numerous solar features, including a thermal storage wall (bottom center), attached sunspace (bottom left), and direct-gain solar (the remainder of the solar glazing). One of the flaws of the design was overglazing: too many windows and skylights in relation to thermal mass. The house tended to overheat year round. Summertime heat gain was especially troublesome because of the numerous skylights and west-facing sliding glass doors (not shown).*

FIGURE I-2

*Trombe walls are one type
of solar design used in homes.
This diagram shows how
they work.*

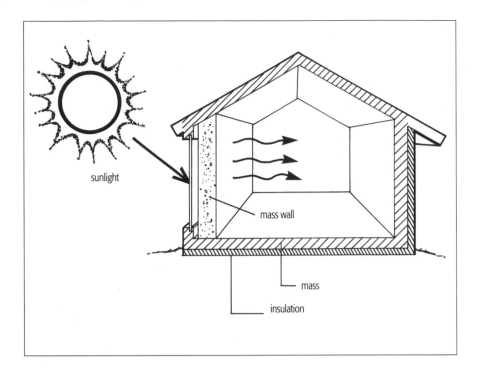

As we drove up the driveway, my first impression was quite favorable. The house was compactly designed, attractive, and located on an extremely nice lot with a spectacular view of Mount Evans, a 14,000-foot peak.

As we stepped out of the car, the main solar feature came into view: a huge thermal storage wall, known as a Trombe wall. Described in more detail in chapter 3, a Trombe wall is typically located on the south side of a home, constructed from concrete block, and covered by glass. Low-angled winter sunlight penetrating the glass heats the wall, and the heat is gradually released into the home (figure I-2). (It moves inward by conduction, flowing from warm to cold, as heat always does.)

We looked around inside, talked with the owners, then toured the barn and walked the property. After looking over the house a second time with my enthusiasm growing, the owner made a statement that sent up a red flag. He mentioned in passing that after waking that morning, he had needed to start a fire in the wood stove to warm up the house. The interior temperature, he told us, had dropped into the low 60s during the night. In fact, he went on to say, he had to heat the home every morning during the heating season, which here consisted of the fall, winter, and much of the spring.

This seemed rather odd to me—all the more peculiar because we were experiencing some rather unseasonably warm and sunny weather. Moreover, in my own passive solar home we hadn't needed any back-up heat that year.

We looked around some more. I noticed that the house we were inspecting was well insulated and the thermal storage wall seemed to be correctly designed. So, as my friend chatted further with the owner, I went outside to contemplate this conundrum. As I stood alongside the house, the source of my puzzlement—and this house's less-than-optimal performance—suddenly became evident: The architect had oriented the home to the west, so the thermal storage wall faced west instead of south. Ideally, a passive solar home should be oriented to the south. This ensures optimal solar gain during the heating season, that is, optimal absorption of sunlight when heat is required.

By orienting the thermal storage wall along the west side of the house, the architect had severely limited solar gain. The wall would not begin to absorb sunlight energy and generate heat until the afternoon. By then, the solar home had missed most of the sun's daily energy contribution. Making matters worse, the south side of the house—the solar side—was an ordinary stud-frame wall with wood siding. It didn't have a single window to permit sunlight to enter the interior of the structure. No wonder the home performed so poorly!

This home is just one of dozens of problematic passive solar structures I've toured and studied since the mid-1970s when my interest in all aspects of solar energy began. Some of the homes I have explored displayed only minor design problems; others, like this one, had major flaws that seriously compromised the thermal performance and overall comfort (figure I-3).

This book is a byproduct of my experience with passive solar design. It is my attempt to set the record straight on passive solar, to outline successful solar design features, to highlight the need for region-specific designs, and ultimately to promote widespread adoption of this exciting and beneficial building technology.

FIGURE I-3
*Like many early passive solar homes, this one suffered from overglazing—too many windows. Early designers reasoned that if a little solar glazing was good, a lot would be even better. Unfortunately, if glazing is not carefully balanced by thermal mass, excess south-facing windows can result in overheating. Overglazing also results in a problem called sun-drenching, described in the text.*

This book also stems from a realization that achieving comfort in our homes takes a huge toll on our personal finances. Americans spend about 54 billion dollars each year to heat and cool their homes. A good portion of our lives is spent working to pay this cost. According to my calculations, a couple earning $60,000 and spending $3,000 on energy each year spends eight hours a month working just to pay their energy bills. We work hard for our homes when, in fact, they should be working for us if at all possible, providing comfort free of charge.

Heating and cooling also incur huge environmental costs. Consuming about one-fifth of the nation's total fossil fuel energy production, our homes contribute to a number of local environmental problems, among them habitat destruction and air pollution. Although most of us attribute local problems such as urban air pollution to the most visible polluters, the automobile and factories, our homes are significant sources as well. Home heating and cooling is also responsible for major regional problems, such as acid rain. Even global environmental travesties can be traced in part to the satisfaction of our ever-escalating requirements for home comfort. According to various estimates, residential heating and cooling contributes about one-fifth of the United States' annual emissions of carbon dioxide gas. This pollutant, while nontoxic, is a principal agent of global warming and climate change. Combined with other problems such as overpopulation and species extinction, global climate change could spawn a future plagued with tremendous ecological and personal misfortune.

As with countless other human activities, the pursuit of our own personal creature comforts rarely takes into account the comfort and well-being of the millions of species that share this planet with us. Many forms of life perish as we live in contentment. Ironically, however, achieving comfort in our homes also occurs at the detriment of our own long-term future. As we produce and consume fossil fuels to meet our immediate needs, we pollute the air and water and deface the landscape, destroying ecological systems that are the life-support system of the planet. What is more, we pursue our comfort virtually unaware of the dangerous social, economic, and ecological backlashes we are propagating.

## NATURAL CONDITIONING: SENSIBLE, ELEGANTLY SIMPLE, AND ECONOMICAL

Fortunately, there are simple and highly affordable ways of ensuring comfortable living spaces that enhance humankind's long-term possibilities and avoid damaging the life-support systems of the planet. These strategies fall within the realm of the frequently overlooked but age-old practice of natural conditioning.

Natural conditioning is the art and science of heating, cooling, lighting, and ventilating a building without outside fuels. It includes four basic, often interrelated, strategies: (1) passive solar heating, (2) passive cooling, (3) daylighting, and (4) natural ventilation.

*We humans often pursue comfort in a moral or ethical vacuum, unaware of the dangerous backlashes we create.*

Whereas modern means of providing comfort typically rely on nuclear and fossil-fuel energy and extraordinarily sophisticated technologies, natural conditioning relies on clean, renewable energy in the form of sunlight and on other natural processes working in concert with sensible building designs. Unlike conventional heating and cooling strategies, natural conditioning achieves comfort through a knowledge of local climate and vernacular design, allowing us to utilize natural elements to provide the amenities we need—lighting, fresh air, and thermal comfort—without bankrupting the Earth. In short, natural conditioning relies on climate-specific or climate-sensitive design (figure I-4).

As you will soon discover, most natural conditioning strategies are alluringly elegant and simple. Their impact on our wallets is negligible, and in many cases beneficial. In fact, climate-sensitive design can save tens of thousands of dollars over the lifetime of a house. Moreover, small investments in knowledge, design, and materials can go a long way toward making our homes comfortable while emancipating society from its costly dependence on fossil fuels.

Successful passive solar heating and cooling does not require a degree in engineering. The Anasazi Indians proved this over a thousand years ago. These ancient people built stone and mud dwellings in the deeply carved canyons of the desert Southwest. Nestled into south-facing canyon walls under natural overhangs, their homes were sheltered from the intense summer sun. Yet as winter approached, the low-angled sunlight dropped below the overhang to provide warmth (figure I-5).

Long before the Anasazi began to take comfort from passive conditioning, across the vast Atlantic, the ancient Greeks were developing ways to utilize the sun's energy to heat their homes. So advanced were the Greeks in their understanding of the importance of sunlight as a source of heat that they treated

*Natural conditioning relies on centuries-old techniques that can produce comfortable homes in virtually every climate where humans dwell.*

FIGURE I-4
*The author's current solar home balances south-facing glass with thermal mass and performs much better than his first home (shown in figure I-1). This house is built from rammed earth tires and straw bales and many recycled building materials. It relies almost entirely on passive solar heating and passive cooling. Note the solar modules on the roof for electricity.*

Catherine Wanek

FIGURE 1-5

*The Anasazi built elaborate dwellings in the deeply carved canyons of the desert Southwest that were heated by the sun.*

## THE ECONOMICS OF PASSIVE SOLAR ENERGY

For those skeptical about the economics of passive solar, numerous detailed case studies demonstrate the point, among them reports published by the American Solar Energy Society in *Solar Today* and the book *Buildings for a Sustainable America: Case Studies* by Burke Miller. Many of these case studies are also available online at a Web site sponsored by the U.S. Department of Energy's Office of Building Technology, www.even.doe.gov/buildings/case_study/.

solar access as a legal right. At least one Greek city, Olynthus, was laid out so that homes would have unfettered access to the sun—and this was in the 5th century B.C.

Not only did the Greeks consider passive solar heating to be a legal right, they deemed it a sign of truly civilized people. They considered those who didn't use passive solar energy to be barbarians!

These and other tried-and-tested methods of natural conditioning were gradually pushed aside in the fossil-fuel age, as elaborate heating systems and air conditioners gained enthusiasts and became the norm. That said, it is important to note that natural conditioning is still being used, quite successfully, in thousands of buildings worldwide. Many new structures in the United States and other industrial nations are also being heated, cooled, lighted, and ventilated by natural means, among them homes, banks, corporate offices, schools, and universities. Not only are these buildings achieving dramatic reductions in fossil-fuel use and energy costs, they often create a level of comfort far superior to that found in many conventionally conditioned structures. In naturally conditioned homes, for instance, the rattling and buzzing of fans and motors to move warm and cool air around are absent. Gone, too, are the annoying and uncomfortable streams of hot or cold air from wall registers characteristic of conventionally conditioned homes. Instead, heat moves passively, radiating out from walls and other structures, creating a more even comfort. By replacing energy-intensive comfort strategies with intelligent building designs that rely on passive conditioning, modern architects and engineers are showing that can we live well while living lightly.

This book focuses on two elements of natural conditioning, passive solar heating and passive cooling, primarily in residential structures. The term *pas-*

*sive* refers to nonmechanical systems of heating and cooling that rely on an assortment of natural forces, such as sunlight in the case of passive heating or cool breezes and openable windows in the case of passive cooling. (You'll learn more about the intricacies of these systems in upcoming chapters.)

Because natural ventilation is a key element of passive cooling, we'll examine this component of natural conditioning as well. We'll also consider daylighting, a beneficial spin-off of passive solar design.

## HOW THIS BOOK IS ORGANIZED

*The Solar House* begins with an overview of passive design principles, a dozen or so guidelines for creating a passively heated/passively cooled home. Chapter 2 examines key aspects of energy-efficient design, essential to passive heating and cooling. Chapter 3 tackles region specific passive solar design. It explores the pros and cons of each major design strategy, and then examines ways to utilize each design to optimal benefit in different regions of the world, that is, in relation to climate and solar (sunlight) availability. Chapter 4 discusses an important, but frequently overlooked, aspect of passive solar design: providing back-up heat in ways that have little impact on the environment. In chapter 5 we turn our attention to passive solar cooling. I outline basic techniques, then discuss effective design strategies for various climates. In this discussion, you'll see that many features that contribute to the success of passive solar heating also contribute to passive cooling.

With the design work required to build a passive solar home clearly outlined, chapter 6 looks at to another extremely important subject: how to maintain indoor air quality in an air-tight, energy-efficient, passively conditioned home. Chapter 7 provides a step-by-step overview of the process of designing and building a passively conditioned home. I also discuss some key tools that enable you to implement an integrated or holistic design process that is crucial to the success of a naturally conditioned home.

Recognizing that to be genuinely sustainable a home requires more—much more—than passive conditioning, in chapter 8, I provide a brief summary of a host of ideas that can help you create a home that contributes to an enduring human presence on the planet. This chapter describes strategies such as solar electricity, graywater and catchwater systems, natural building technologies, and green building products—all vital to the quest for sustainable living.

## SOLAR DISTINCTIONS

Solar energy systems come in several different forms, each of which performs a unique function. The three most commonly encountered systems are: (1) passive solar heating, (2) solar hot water, and (2) solar electric.

**Passive solar heating systems** are designed to provide space heat by capturing sunlight through solar-oriented windows. The sun's energy is converted to heat, which warms the interiors of buildings.

**Solar hot water systems** are designed to produce hot water for domestic uses, such as washing dishes or showering, and/or heating interior spaces. These systems rely on collectors, often mounted on rooftops, that absorb sunlight energy and transfer the heat to a fluid that transfers the heat to water.

**Solar electric systems** are designed to produce electricity, usually with the aid of special panels called modules that contain a semiconductor material that generates a flow of electrons when struck by sunlight. These are also known as *photovoltaic systems.*

*With foresight, careful planning, knowledge, and common sense we can achieve comfort in virtually any climate naturally using sunlight, shading, earth sheltering, insulation, and natural daylight.*

# 1 FUNDAMENTALS OF INTEGRATED PASSIVE DESIGN

**ONE OF THE KEYS** to successful passive conditioning is *integrated design:* designing buildings as whole, functional units. The goal of this approach, also called holistic design, is to achieve year-round comfort with minimal impact and cost.

Integrated design requires an understanding of the relationships between building components—for example, the relationship between south-facing windows (solar glazing) and heat-absorbing mass (thermal mass) in a home. Integrated design also requires an understanding of the ways in which design elements and various building materials contribute to and detract from various goals. Moreover, this exciting new approach to building design and construction calls for a detailed understanding of the ways that design features incorporated to achieve one goal may conflict with other essential objectives. For example, skylights permit additional sunlight to enter a home during the heating season, thus contributing to passive solar heating. However, skylights can permit excess solar gain in periods when heating is not required—the summer, for instance. This beneficial feature, therefore, strains the capacity of the passive cooling design elements. Skylights may also cause excessive heat loss at night during winter months, also reducing their overall value. (Such contraindications may lead an architect to eliminate a design element.)

Integrated design seeks multiple benefits from each and every design feature. South-facing windows, for example, permit passive solar gain, but also provide light during the day. Daylighting reduces the need for artificial light and thus saves energy and money while lessening environmental impacts. In addition, south-facing glass allows occupants generous views of their surroundings. Views, in turn, often add to the aesthetic value of a home and personal satisfaction. They may also boost the resale value of a passively conditioned home.

## INTEGRATED DESIGN

Integrated or holistic design requires an understanding of the interaction of building components to achieve a home whose parts function well together. Two of the benefits are increased comfort and reduced energy consumption.

The key to successful passive solar design is to think about a home as a package of features that work optimally together to provide heat when you need it, cool indoor temperatures when it is hot outside, and sufficient daylight and fresh air for people to remain healthy and productive. The overall effect is a consistently pleasant indoor environment and a home with minimal dependence on supplemental energy, furnaces, air conditioners, and fans.

Integrated design frequently requires a collaborative effort involving owners, builders, architects, and subcontractors, especially heating and cooling professionals. Meetings among participants permit the free exchange of ideas and continual attention to detail and relationships that are crucial to designing and building a structure that functions as an efficient, integrated whole. Of course, efforts to work together must start early in a project cycle.

Granted, the team approach may take longer and may cost more up front, but the process often results in a far superior building than is possible when relying on the more traditional approach to design and construction. In this often-fragmented approach, the parts of a building project are typically divided up among different parties. The architect designs the home, the builder builds it according to the architect's specifications, and an HVAC contractor installs the heating, ventilation, and air conditioning system. Communication between the participants is often limited, except on matters of schedule. The result is often a house that is poorly equipped to take advantage of passive solar heating and cooling. Energy, money, and elaborate systems are required to maintain comfort. Muscle, rather than intelligence, is the byword. In sharp contrast, the team approach helps to ensure the incorporation and integration of the elements needed to create a passively conditioned home and increases the chances that savings from passive solar heating strategies are not lost through mediocre walls, roofs, and floors.

The integrated design process is well-suited not only to the professional building community but also to owner-builders, who often engage spouses, friends, and family as team members. Whether you are working with a team of professionals or a group of friends and relatives, integrated passive design demands an understanding of key principles. We'll examine fourteen principles of integrated passive design in this chapter, which inform and direct the process from the outset. As you study these guidelines, bear in mind that most principles are complementary: They contribute to passive heating and passive cooling. Within this discussion, I will also present specific details vital to the siting and design of a passively conditioned home. Many more details will follow in subsequent chapters.

## PRINCIPLE 1: CHOOSE A SITE WITH GOOD SOLAR EXPOSURE

Passive solar requires sufficient sunlight during the heating season, beginning with unobstructed access to the sun, ideally from 9 AM to 3 PM or at least 10 AM to 2 PM, during the heating season. In colder climates, the heating season usually occurs in the late fall, winter, and early spring. In warmer climates, the heating season is substantially shorter. In southern Florida, it may last a month or so.

Simple as this fundamental requirement sounds, many important factors come into play. To begin, consider the most obvious issue, the amount of sunlight or solar radiation falling on a region. The map in figure 1-1 shows solar radiation by region. As a rule of thumb, the sunnier the area, the better the prospects for passive solar heating.

Contrary to popular belief, however, passive solar heating is suitable in many less sunny areas as well. In fact, most parts of the United States receive enough solar energy during the heating season to merit an investment in passive solar design.

Passive solar heating works even in problematic climates. In the cloudy, snowy northeastern United States, for example, passive solar can provide a significant amount of heat, not so much during the dead of winter, but in the months before and after the cold, snowy weather arrives. Consider Buffalo, New York, located in what some designers refer to as the "gloom belt." Buffalo receives very little sunshine during the winter months. In December, the sun shines only about 26 percent of the available daylight hours. During January that figure climbs to a paltry 32 percent, and during February it increases to 38 percent. During this period, then, solar energy provides only a modest amount of heat. However, during the three months before and three months after the gloomy season, the so-called "swing season," ample sun is generally available to heat a home. With careful design and attention to construction and materials, an architect can design a home in Buffalo that will satisfy over 50 percent of its annual heat demand from solar energy.

Therefore, while it is important to know the amount of sunlight striking an area each year, a knowledge of seasonal patterns is equally vital. Figure 1-1 shows the seasonal availability of sunlight in a few select locations. Appendix A lists monthly data for selected cities in the United States and Canada. (If you're from Europe, Australia, or elsewhere, check with your nation's national weather agency for data on your location.)

## Sunrise, Sunset: Bearing Angle

Knowing annual solar radiation (insolation) and seasonal variations is important, indeed essential. But how do you know if a specific site will work for passive solar design?

The answer requires a good working knowledge of the sun's path through the sky at various times of the year. You'll need to know two things: (1) where the sun rises and sets, and (2) the arc the sun "cuts" through the sky during each season.

As any school child should know, the sun rises in the east and sets in the west. However, if you've studied the sun's path for any length of time, you

*Contrary to popular belief, passive solar heating is not restricted to the sunniest locales on our planet.*

FIGURE 1-1
*Annual mean daily solar radiation and mean percentage of possible sunshine by month in selected locations.*

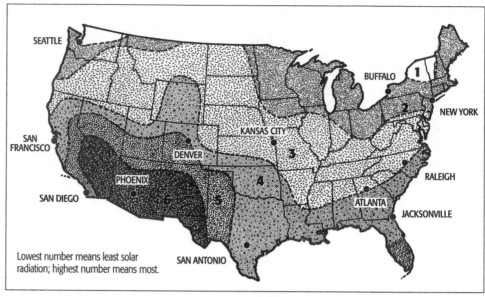

Lowest number means least solar radiation; highest number means most.

|  | JANUARY | FEBRUARY | MARCH | APRIL | MAY | JUNE | JULY | AUGUST | SEPTEMBER | OCTOBER | NOVEMBER | DECEMBER |
|---|---|---|---|---|---|---|---|---|---|---|---|---|
| Buffalo | 32 | 41 | 49 | 51 | 59 | 67 | 70 | 67 | 60 | 51 | 31 | 28 |
| New York | 49 | 56 | 57 | 59 | 62 | 65 | 66 | 64 | 64 | 61 | 53 | 50 |
| Raleigh | 50 | 56 | 59 | 64 | 67 | 65 | 62 | 62 | 63 | 64 | 62 | 52 |
| Atlanta | 48 | 53 | 57 | 65 | 68 | 68 | 62 | 63 | 65 | 67 | 60 | 47 |
| Jacksonville | 58 | 59 | 66 | 71 | 71 | 63 | 62 | 63 | 58 | 58 | 61 | 53 |
| Kansas City | 55 | 57 | 59 | 60 | 64 | 70 | 76 | 73 | 70 | 67 | 59 | 52 |
| Denver | 67 | 67 | 65 | 63 | 61 | 69 | 68 | 68 | 71 | 71 | 67 | 65 |
| San Antonio | 48 | 51 | 56 | 58 | 60 | 69 | 74 | 75 | 69 | 67 | 55 | 49 |
| Phoenix | 76 | 79 | 83 | 88 | 93 | 94 | 84 | 84 | 89 | 88 | 84 | 77 |
| San Diego | 68 | 67 | 68 | 66 | 60 | 60 | 67 | 70 | 70 | 70 | 76 | 71 |
| San Francisco | 53 | 57 | 63 | 69 | 70 | 75 | 68 | 63 | 70 | 70 | 62 | 54 |
| Seattle | 27 | 34 | 42 | 48 | 53 | 48 | 62 | 56 | 53 | 36 | 28 | 24 |

## BEARING ANGLE

The bearing angle is the position of the sun at different times of the day in relation to true north. True north is a line that runs from the South Pole to the North Pole. It differs from magnetic north, determined by a compass, because magnetic fields rarely run true north and south.

know that the location of sunrise and sunset changes daily. During winter months in Colorado, for instance, the sun rises in the southeast and sets in the southwest. As days get longer, the sunrise occurs farther and farther to the north. Sunset shifts northward as well. By the time summer rolls around, the sun is rising in the northeast and setting in the northwest. By June 21, the longest day of the year, the sun reaches its farthest northern position at both sunrise and sunset. Then, over the next six months, from June 22 until December 22, the sun retraces its steps, rising farther and farther to the southeast and setting a little closer to the southwest each day.

Figure 1-2 illustrates the position of the sun at sunrise and sunset at various times of the year in Denver, Colorado. The dashed line in the drawing shows the position of sunrise and sunset in relation to true north, known as the *bearing angle*. Take a moment to study the diagram. Note where the sun rises and sets on each of the following days: December 22 (winter solstice), March 22 (spring equinox), June 22 (summer solstice) and September 22 (fall equinox).

The angle of the sun in relation to true north also changes during the day as the sun courses across the sky (actually as the Earth rotates). The numbers in circles in the illustration represent the time of day.

An understanding of bearing angle is useful primarily because it helps a designer determine the best site for a home on a particular piece of property in relation to potential obstructions. A grove of trees, for instance, may block the low-angled sun from 10 AM to 12 noon. Because that is a significant portion of

**FIGURE 1-2**

*This drawing shows the bearing angle, the relation of the sun to true north at different times of the year and different times of the day. The insert shows the altitude angle—the angle of the sun from the ground—at the equinoxes and solstices.*

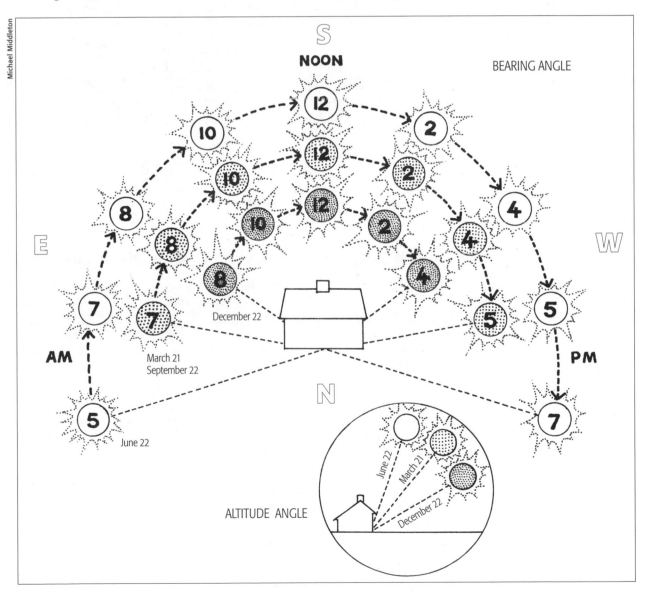

the daily heating period, you would likely want to try a different location. If the trees blocked the sun after 4 PM, less sun will be lost; most solar heating has already occurred by this time.

To determine the bearing angle on a piece of property or a city lot, you will need to know the latitude of the site. Figure 1-3 is a map of the United States and Canada showing latitude. Find your location, then consult a chart of bearing angles for your region.

**FIGURE 1-3**

*The altitude and bearing angles of the sun vary with the latitude. Consulting a chart, like the one shown here for latitude 40° north, provides detailed information on both altitude and bearing angle.*

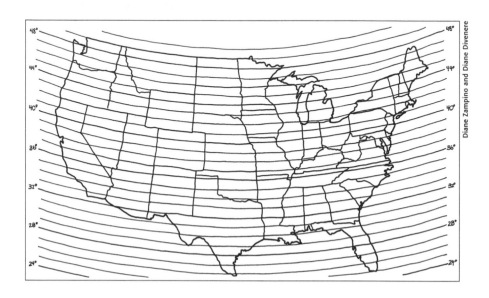

## ALTITUDE ANGLE AND BEARING ANGLE AT 40° N. LATITUDE

| | JUNE 22 | | | MARCH 21, SEPTEMBER 22 | | | DECEMBER 21 | |
| TIME | ALTITUDE ANGLE | BEARING ANGLE | TIME | ALTITUDE ANGLE | BEARING ANGLE | TIME | ALTITUDE ANGLE | BEARING ANGLE |
|---|---|---|---|---|---|---|---|---|
| 5:00 AM | 4.23 | 62.69 | 6:00 AM | 0 | 90.00 | 8:00 AM | 5.48 | 127.04 |
| 6:00 AM | 14.82 | 71.62 | 7:00 AM | 11.17 | 100.08 | 9:00 AM | 13.95 | 138.05 |
| 7:00 AM | 25.95 | 80.25 | 8:00 AM | 22.24 | 110.67 | 10:00 AM | 20.66 | 150.64 |
| 8:00 AM | 37.38 | 89.28 | 9:00 AM | 32.48 | 123.04 | 11:00 AM | 25.03 | 164.80 |
| 9:00 AM | 48.82 | 99.81 | 10:00 AM | 41.21 | 138.34 | 12:00 PM | 26.55 | 180.00 |
| 10:00 AM | 59.81 | 114.17 | 11:00 AM | 47.34 | 157.54 | 1:00 PM | 25.03 | 195.19 |
| 11:00 AM | 69.16 | 138.11 | 12:00 PM | 49.59 | 180.00 | 2:00 PM | 20.66 | 209.35 |
| 12:00 PM | 73.44 | 180.00 | 1:00 PM | 47.34 | 202.45 | 3:00 PM | 13.95 | 221.94 |
| 1:00 PM | 69.16 | 221.88 | 2:00 PM | 41.21 | 221.65 | 4:00 PM | 5.48 | 232.95 |
| 2:00 PM | 59.81 | 245.82 | 3:00 PM | 32.48 | 236.95 | | | |
| 3:00 PM | 48.82 | 260.19 | 4:00 PM | 22.24 | 249.32 | | | |
| 4:00 PM | 37.38 | 270.71 | 5:00 PM | 11.17 | 259.91 | | | |
| 5:00 PM | 25.95 | 279.74 | 6:00 PM | 0 | 270.00 | | | |
| 6:00 PM | 14.82 | 288.37 | | | | | | |
| 7:00 PM | 4.23 | 297.30 | | | | | | |

The bottom chart in figure 1-3 shows the bearing angle at 40° latitude. Notice that the chart lists June 22, March 21 and September 22, and December 21. The bearing angle at sunrise for each of the listed days is the top number in the columns labeled "bearing angle." The bottom number in that column is the position of the sun at sunset. In between are the bearing angles at different times of day. You'll note that the columns vary in length. That's because day length varies from season to season.

You can find a chart of bearing and altitude angles for many latitudes at my Web site, listed in the Resource Guide. If your site isn't included, you may be able to obtain a chart through a solar supplier or your local library. A meteorologist at a local TV station may have a chart or may be able to tell you where you can locate one. Bearing angle data can also be found on the Internet, at a Web site called Sun Angle (www.sundesign.com/ sunangle). It provides access to an on-line program that calculates bearing angles at different times of the year and different times of the day. All you need to do is to provide the latitude and longitude of your site.

## Altitude Angle

After you have determined the seasonal bearing angle for your site, you will need to determine the altitude angle: the angle of the sun from the ground at crucial times of the year (figure 1-2, inset). The altitude angle, like the bearing angle, is useful for siting a home to avoid obstructions. It also affects design

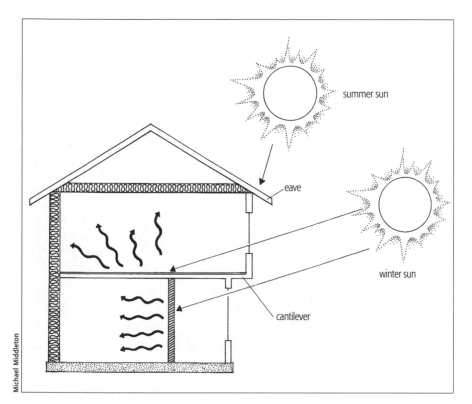

Michael Middleton

FIGURE 1-4
*Sunlight streams through windows when the sun is low in the sky during the late fall, winter, and early spring. As the sun "rises" in the sky, less and less sunlight penetrates the windows, automatically shutting off the heat source that is no longer needed.*

Trees can be a valuable ally in passive solar designs. Planted on the east and west sides of a home, they can block low-angled sun that may contribute to summertime overheating in many climates. Even small fruit trees can shade windows and walls.

elements such as overhang and mass placement, required to ensure maximum year-round performance of a passively conditioned home.

Like bearing angle, the altitude angle changes from one day to the next over six-month periods. Let's begin with December 22, the shortest day of the year. On this day, also known as the *winter solstice,* the sun traces its lowest arc across the sky. In a passive solar home, the low-angled sun penetrates south-facing windows, casting light into the house. Sunlight is then converted into heat (figure 1-4).

Six months later, on June 22, the longest day of the year, the sun is at its highest point in the sky (figure 1-4). It tends to shine on the roof of a home, not the south-facing walls. The result: very little solar gain.

If the shortest day of the year is December 22 and the longest is June 22, half way between each extreme are days that have equal hours of daylight and darkness. These are known as *fall* and *spring equinoxes* (equinox means equal night). The spring equinox occurs on March 22 and the fall equinox occurs on September 23. On the equinoxes, the sun is in an intermediate position, providing some solar gain.

## Unobstructed Access to the Sun

Ensuring unobstructed access requires careful site selection. Open sites are best. Avoid heavily wooded lots, or plan to remove some trees on the sunny side of the house.

This doesn't mean that solar sites must be devoid of trees. A few deciduous trees around a house are fine, sometimes beneficial, as they shade a house in the summer, and thus contribute to passive cooling. Far more important are the location of trees in relation to the house and the species. Let's consider location first: Trees generally do not pose a problem as long as they are located on the north, west, and east sides of the home, ensuring unobstructed access to

**FIGURE 1-5**
*These trees were small when the house was built over twenty years ago. They have now grown up and block much of the sunlight from this passive solar home.*

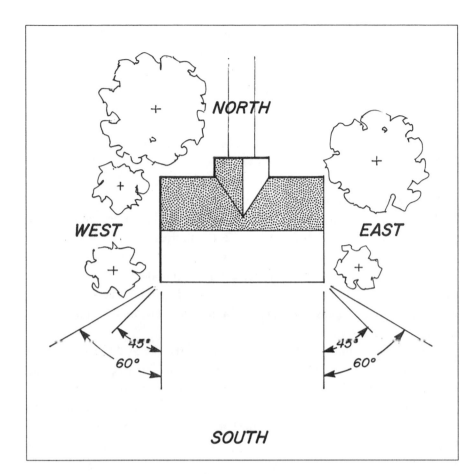

NORTH

WEST

EAST

45°

60°

45°

60°

SOUTH

FIGURE 1-6
*Ensuring unobstructed access to the sun from 9 AM to 3 PM (60°-zone) or 9 AM to 2 PM (45°-zone) is essential for passive solar heating. Trees on the north, east, and west sides generally cause no problems, and in fact may be beneficial in protecting a home from winter winds. Trees on the east and west sides help shade a home during the summer.*

the sun during the heating period. Some deciduous trees along the south side may be acceptable. However, not all deciduous trees are created equal. Oaks, for example, retain dead leaves throughout part of the fall and winter and thus can reduce solar gain during part of the heating season. Even deciduous trees that give up their leaves more readily than oaks can block full access to winter sun, as their trunks and limbs obstruct the sun. As a general rule, unless decid-uous trees on the south side of a passive solar home are located near the house and lower branches are removed, they will reduce solar gain.

Evergreen trees on the south side of a home are even more problematic (figure 1-5). If located too close to the house, they can block the low-angled winter sun completely. According to the Sustainable Buildings Industry Coun-cil, evergreens should be located at least three times their projected (mature) height away from the south wall of your house. For example, if an evergreen tree will reach 50 feet at maturity, it should be about 150 feet from the south side of a house to prevent shading during the winter when the sun is at its low-est point in the sky.

Mountains, hills, cliffs, or other buildings can also block solar radiation. Potential obstructions should not be located within an arc of 60 degrees on either side of true south, which corresponds to the 9 AM to 3 PM solar window

*Your main objective in siting a passive solar home is to ensure the south side obtains as much unshaded exposure as possible during the heating season.*

**FIGURE 1-7**
*This device, known as the Solar Pathfinder, sizes up potential obstructions and assesses the suitability of a site for passive solar heating quickly and accurately.*

courtesy Solar Pathfinder

*A building elongated on an east-west axis inherently captures more solar radiation in winter and reduces the potential for unwanted summer gain.*

Designing Low-Energy Buildings, Sustainable Buildings Industry Council

mentioned earlier. Good solar access is still possible if the glazing is unshaded with an arc of 45 degrees or from 10 AM to 2 PM (figure 1-6).

Fences and buildings can also obstruct solar radiation. Spacing for fences and other buildings varies with latitude. The higher the sun is on December 22, the closer a structure can be located. In Denver, located at about 40° latitude, no structures should be closer than 13 feet. Fences are permissible in a 13-to-23 foot zone. One-story buildings can be within 23 to 53 feet. Two-story buildings should be at least 53 feet away.

## Assessing the Suitability of a Site

With a compass and charts in hand, you should be able to determine where sunrise and sunset occur on December 22, the shortest day of the year when the sun is lowest in the sky. You should also be able to locate sunrise and sunset on June 22, the longest day of the year. In the middle lies the sunrise and sunset positions on March 22 and September 22. This information, combined with altitude angle data, should enable you to determine the sun's path across the sky at any given site.

Knowledge of the sun's "movement" across the sky at different times of year allows you to determine the suitability of a site—that is, whether a site will provide good solar access. Trace the sun's path and see how obstructions, if any, will block it. Remember: your main objective is to ensure that the south side of the building receives as much unshaded exposure as possible during the heating season.

## Hire a Professional or Use Solar Pathfinder

When in doubt, you can hire a solar designer or solar builder to check out a site or review your evaluation. Although it may cost you a little money to hire a professional, the expense is well worth it.

Another form of assistance is the Solar Pathfinder (figure 1-7). This handy piece of equipment analyzes the solar suitability of any site. After you set it up, one quick measurement by Solar Pathfinder gives accurate solar information for the entire year. How does it work?

Solar Pathfinder views all potential obstructions that could block solar access. Armed with this information and programmed with site-specific data on bearing and altitude angles, this marvelous little piece of technology determines solar availability and potential energy loss resulting from various obstructions. Solar Pathfinder is especially valuable when comparing two or more building sites.

## Wind Flow and South-Facing Slopes

While you are studying the path of the sun across the building site, you should also take note of the wind direction. If you are in a colder climate, you will want to take precautions to protect your house from wind, which can rob heat from buildings. Ask knowledgeable neighbors and friends which direction the winds blow; usually the direction corresponds with the prevailing movement of storms through the area, although wind patterns can change from one season to the next and odd wind patterns develop in valleys and other locations.

Building a house on a south-facing slope will enhance its potential for passive heating, as south-facing slopes tend to be warmer than other locations and may shield a home from cold winter winds. The result is a warmer microclimate with lower heating demands.

## PRINCIPLE 2: ORIENT THE EAST-WEST AXIS OF A HOUSE WITHIN 10 DEGREES EAST OR WEST OF TRUE SOUTH

For optimal performance, a solar home needs to be oriented properly on a site to obtain as much sunlight as possible. The rule of thumb for best solar gain is

FIGURE 1-8
*For ideal solar gain, orient a home so its long axis is aligned with true east and west. This places the maximum amount of south-facing wall toward true south.*

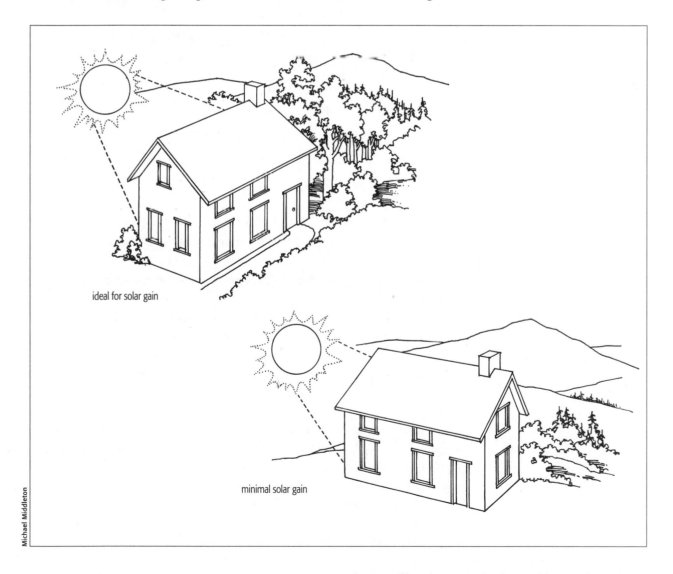

ideal for solar gain

minimal solar gain

Michael Middleton

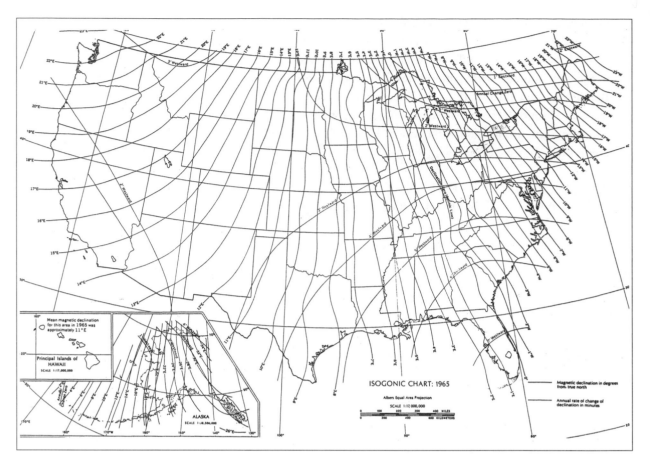

**FIGURE 1-9**

*True north and south rarely correspond with magnetic north and south, except in the midsection of the country.*

that the long axis (east-west axis) of the house should be oriented within 10 degrees of true south (figure 1-8). Orienting a house to the south presents the largest possible amount of exterior wall space to the sun. Because the south-facing wall is fitted with windows, this orientation ensures maximum winter-time solar gain.

Note that the description of orientation just given requires alignment of a home in relation to true south. True north and south follow a line that runs directly from the North to the South Poles. They are geographic terms and not the same as magnetic north and south. Magnetic north and south are deter-mined by magnetic fields created by iron-bearing minerals in the Earth's crust and determined by a compass. In most locations true north and south and magnetic north and south do not coincide (figure 1-9). In El Paso, Texas, for example, true south is actually 12 degrees east of magnetic south. If you live there, take your compass outside. Locate magnetic south. Now draw a line pointing 12 degrees east of this line. This is true south.

Figure 1-9 shows the relation of magnetic south to true south in the United States. Note that in the western portion of the country, magnetic north and south digress quite dramatically from true north and south.

## Deviating from the Rule: Reducing Wintertime Solar Gain, Increasing Summer Gain

Unfortunately, solar homes cannot always be oriented within 10 degrees of true south. Exceptional views, and considerations such as placement of homes in relation to streets, may dictate different alignments. Although deviating from the 10-degree rule reduces solar gain during the heating season, the decline is not noticeable until the deviation is quite significant. As illustrated in figure 1-10, a house oriented directly south gains 100 percent of the solar energy it can absorb. At 22.5 degrees east or west of true south, solar gain falls slightly, to 92 percent. At 45 degrees, however, potential solar gain plummets to 70 percent.

Even though departure from the 10-degree rule may cause only a small decrease in solar heat gain during the winter, it can exact a huge penalty in the summer by increasing summertime heat gain—a fact that underscores the complementarity of passive solar design principles. In warm climates, or when deviation is large, overheating is very likely. That's because any deviation from true south shifts the orientation of the east and west sides of the house more toward the summer position (bearing angle) of the sun early and late in the day. Sunlight striking the east and west walls and windows can lead to excessive heat gain—and burning indoor temperatures. Deviation from the 10°-rule may also result in significant solar gain in solar glazing early or late in the day.

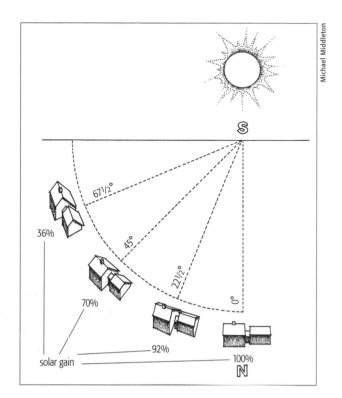

**FIGURE 1-10**

*The ideal orientation for a passively conditioned solar home is with its long axis perpendicular to true south (0° on the diagram).*

**FIGURE 1-11**

*While north-south orientation of homes often allows for better alignment to streets and views, it reduces solar gain and increases cooling loads. An L-shaped design allows access to views and permits appropriate alignment to streets, but increases the southern exposure, thus increasing solar gain.*

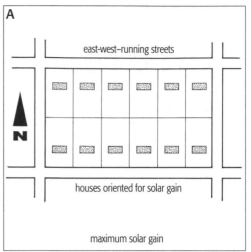

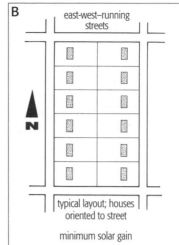

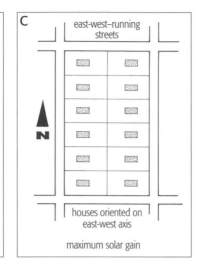

**FIGURE 1-12**

*With little effort, designers of housing developments can orient homes for optimal solar gain. Developers and builders who are knowledgeable about the benefits of passive solar design have found many ways to tap into the sun's generous supply of energy. (a) In subdivisions in which the streets run east to west, houses can be oriented on an east-west axis, with windows concentrated on the south side. (b) Passive solar can be incorporated in houses in subdivisions with north-south running streets, as well, by concentrating glass on the south walls of homes oriented toward the streets—that is, on a north-south axis. (c) Alternatively, the long axis of the houses can be shifted to an east-west orientation. There is no excuse for not incorporating at least some passive solar!*

The practical upshot of this is this: The colder the climate—or the more heat you want from sunlight—the closer toward true south a home should be aligned on an east-west axis. In warmer climates, you don't have to be so particular about orientation when it comes to wintertime heat, but you will pay a huge penalty in the summer. If you have a choice, a southeast orientation is almost always better than a southwest orientation because summertime solar gain is greatest when the orientation of a building is shifted to the west.

While we're on the subject, even when views and other factors result in a less-than-ideal orientation, simple techniques can enable a designer to capture a sufficient amount of solar energy. For instance, if the dominant view occurs to the east, a home can be oriented with its long axis to true south to capture solar energy so long as there are sufficient windows in locations such as master bedrooms and living rooms that permit access to the view. Better yet, the house can be designed in an L-shape, with the short arm of the L directed toward the vista (figure 1-11). The long arm of the L is oriented on an east-west axis for maximum solar gain. Passive solar heating can even be incorporated into homes in new subdivisions where orientation is often rigidly determined by street orientation, which is seemingly unsuitable for solar (figure 1-12).

## Rectangular Floor Plans

Orienting a house to the south, as noted above, means orienting its long axis east and west, so that the house presents the largest possible amount of exterior wall space and window area to the south. As a general rule, a rectangular floor plan works best for passive solar design. The ideal length to width ratio is 1.3 to 1.5. When properly oriented, it gathers the maximum amount of sunlight from south-facing windows.

Rectangular floor plans also minimize the exposure of east and west walls to summer sun. This is especially important in hot climates as east- and west-facing windows can result in substantial solar gain during the summer, causing mild-to-severe overheating. In such instances, passive cooling measures may become insufficient. To achieve desired comfort, you may need to install an active cooling system, such as a heat pump, a swamp cooler, or an air conditioner.

If you're considering building in such an area and would like a few west-facing windows to take advantage of a view, one option is to partially shade them. Trees, arbors planted with deciduous vines, window shades, or awnings are possible options. However, shading results in a trade-off because most of these options will obstruct the view you were seeking in the first place. Special glazings, described in chapter 2, and window films that reduce solar transmission, may be useful in such situations.

In cooler climates, solar gain from east- and west-facing windows may not be so serious a matter. Study the climate carefully before making a decision that could cost you dearly in the long run.

## PRINCIPLE 3: LOCATE MOST WINDOWS ON THE SOUTH SIDE OF A HOUSE*

Windows and patio doors are solar collectors in a passive solar home. For optimal performance, they must be concentrated on the south side of a house in the northern hemisphere. As noted earlier, south-facing windows warm the interior automatically when heat is needed. When the sun is at its lowest point in the sky on the winter solstice, sunlight sun can penetrate over twenty feet into a house. As the sun rises in the sky, less and less sunlight enters, generating less heat as days grow warmer.

Skylights also serve as solar collectors in passive solar designs when located on south-facing roofs. Although skylights also increase daylighting, they do pose some substantial problems, for example, excessive heat loss in the winter and summertime overheating.

A passive solar home incorporates a sufficient number of windows (and sometimes skylights) for heat and daylighting; however, don't make the common assumption that extraordinary allocations of wall space must be dedicated to solar glazing. As noted earlier, a passive solar home must include enough solar glazing for optimal performance in the winter, but not too much. The amount of solar glazing required in a home depends on the amount of heat one wants to obtain, but it must be carefully scaled to the amount of thermal mass contained in the building. I'll present recommendations in chapter 3.

## PRINCIPLE 4: MINIMIZE WINDOWS ON THE NORTH, WEST, AND EAST SIDES

Passive design requires judicious use of nonsolar windows, that is, glazing on the north, east, and west walls. Too many windows on east and west walls can

---

* In the Northern Hemisphere windows should be concentrated on the south side; in the Southern Hemisphere they should be on the north side.

*Choosing a good building shape and orientation are two of the most critical elements of an integrated design.*

*Designing Low-Energy Buildings*, Sustainable Buildings Industry Council

### LOW-E WINDOWS

Most low-e windows are made by applying a thin, transparent coating of a special material to the inside surface of one of the panes in a double-pane window unit. This coating blocks the transmission of long wavelength infrared radiation, or heat. Low-e windows retard heat loss or gain, depending on where the special coat is applied.

increase summertime heat gain. Too many north-facing windows can cause excessive heat loss in the winter. I'll present specific recommendations for north, east, and west windows in chapter 3. A good knowledge of local conditions is crucial when designing windows in a passive solar home.

## Compensating for Design Compromises

Heating and cooling penalties resulting from a deviation from recommended window placement guidelines can be mitigated in part by installing specially designed windows. As you'll see in chapter 2, manufacturers now produce a large assortment of special-application windows. Some reduce heat flow across the panes of glass; others cut back on solar gain (sunlight penetration). When heat transfer, either into or out of a window, is expected to be problem, be sure to consider windows equipped with low-e glass. Low-e stands for low heat emissivity. Translated, it means low heat transmission.

## PRINCIPLE 5: PROVIDE OVERHANGS AND SHADING TO REGULATE SOLAR GAIN

In passive solar homes, sunlight is a welcome guest during the heating season. It's presence, however, may wear thin during the early fall or spring—seasons when the sun is still fairly low in the sky yet heating demands are reduced. In such instances, low-angled sun penetrating south-facing windows and warm daytime temperatures may cause overheating, unless overhangs are provided to regulate solar gain.

The overhang or eave of a home is a relatively simple design element that determines when sunlight begins to penetrate south-facing windows (when the sun can begin to heat a home) and when sunlight penetration ends (when solar heating ends). In other words, overhangs determine the start-up and cut-off dates of solar gain (figure 1-13). As a result, overhangs are vital to both passive heating and cooling of a home.

Overhangs work by reducing solar gain at intermediate sun positions. They also shade windows and walls in the summer, helping to keep the interior of a home cooler. In addition, overhangs protect walls from driving rain, a feature that is especially important in natural homes, such as those made from straw bales or adobe.

"The optimal size of an overhang on the south side of a building depends on the relative importance of heating and cooling loads, and on the latitude of the site," note the authors of *Designing Low-Energy Buildings*. The greater the heating load, the smaller the overhang.

Designing the proper overhang, however, is often a matter of compromise. In colder climates, for example, the spring tends to be chilly and cold; the fall tends to be sunny and warm. As a result, passively conditioned homes need considerably more solar heat to maintain comfort during the spring than they

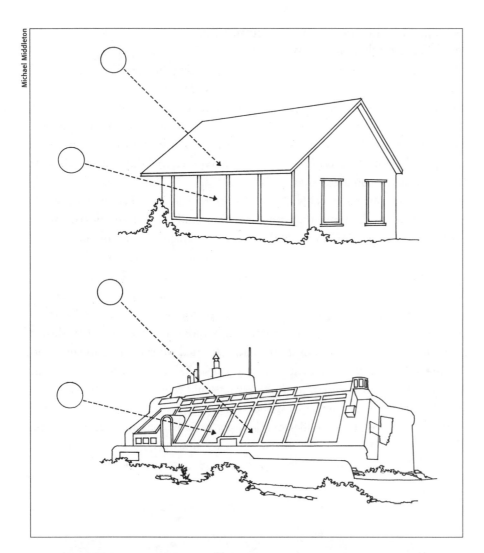

FIG 1-13
*Overhangs regulate solar gain—that is, when sunlight can begin to enter a solar home and when solar gain ends each year.*

do in the fall. However, because the sun is at the same altitude angle both times of the year, a long overhang will block too much sun in the spring when heat is still required. A short overhang would let excessive amounts of sunlight into a house in the fall, when heating is not required.

What designers generally aim for in such instances is a compromise: an intermediate-sized overhang. It permits greater sunlight in the spring, allowing for more solar gain. It also permits greater solar gain in the fall, but the excess is generally reduced by using window shades. When building a home in such a region, you may want to consider a fairly short overhang for optimal solar gain in the spring. You can always use external or internal window shades to block excess solar influx in the fall.

In hotter climates, such as southern California and the southeastern United States, spring and fall are both warm. In such instances, one size serves all. I'll describe how you determine overhang projection in chapter 3.

## Special Requirements for Protecting East- and West-Facing Windows

While overhangs regulate solar gain on south-facing windows, they are generally ineffective on east- and west-facing glass. The reason is that early- and late-day sun is at very low angle in the sky and shines on walls and windows despite the presence of overhangs. A fixed overhang is therefore fairly useless on east- and west-facing walls, except to protect the walls from rain. As a result, some other means of shading may be required to reduce unwanted solar gain during the cooling season.

Trees are often a valuable ally against low-angled sun. Planted on the east and west sides of a home, they block the early and late sun that warms walls and penetrates window glass, both of which may contribute to summertime overheating. Vine-covered arbors and trellises mounted against the side of a house also provide effective shade for east- and west-facing walls and windows.

External window shading is another means of providing protection against unwanted heat gain. Canvas awnings, roll-down blinds, and vertical louvers all prevent sunlight from entering windows. Experience has shown that these are a better option than shades located on the inside of windows (figure 1-14). Internal shades are an option as well, but they're generally not as effective as external shades, for reasons that will become clear in chapter 5.

While shading strategies can be very effective, they do increase initial building costs. Window shades may require periodic maintenance and repair. So be careful. Don't fall into the common trap of overglazing, then shading. Size and locate windows correctly in the first place to prevent overheating.

**FIGURE 1-14**

*External awnings work better than internal shading devices to prevent unwanted solar gain during the cooling season. As you can see from this illustration, many options are available.*

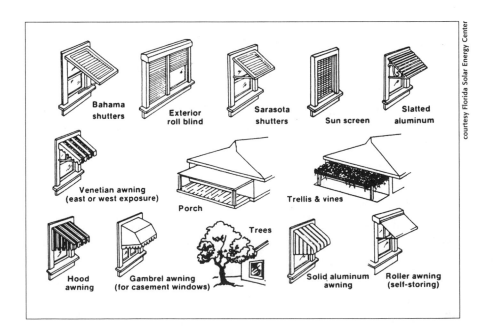

courtesy Florida Solar Energy Center

## PRINCIPLE 6: PROVIDE SUFFICIENT, PROPERLY SITUATED THERMAL MASS

Thermal mass is an important component of all passively conditioned buildings, because it helps to stabilize internal temperatures year-round.

### Free vs. Intentional Mass

Thermal mass is any dense material that absorbs and stores heat. Thermal mass fits into two categories. The first is "free" or "incidental" mass located in standard building components such as drywall, framing lumber, doors, and furniture. Free mass is part of the structure of a house or the room decor. It wasn't installed to function as heat-absorbing mass, but it performs the job anyway.

The second type of thermal mass is intentionally incorporated into home design and strategically placed in the house to absorb sunlight and heat. It is called *intentional mass* (figure 1-15). Mass partition walls are an example.

Brick, concrete, tile, earthen materials, and other dense matter are all suitable forms of thermal mass. They are usually incorporated into the house design to serve other functions as well—tile flooring and brick masonry heaters are good examples. When considering costs, be sure to remember their multiple roles.

To function properly, thermal mass must be dense. Most masonry products used in passive solar homes, such as brick, concrete, and concrete blocks, have similar densities and similar heat-storage properties. Earthen building materials are less dense, but are still sufficient to serve as thermal mass (table 1-1).

### Interior vs. Exterior Mass

In most passively conditioned homes, mass is located within the interior of a well-insulated structure, in floors, partition walls, ceilings, and furnishings. In stick-frame houses, for instance, drywall forms a thin shell of distributed mass. Tile in kitchens, dining rooms, and hallways provides additional mass.

In passive solar homes, designers generally seek to deliver sunlight into the interior of the structure, where it is absorbed by internal mass. Envelope insulation—insulation in the walls, floors, ceilings, and foundation—helps to preserve solar heat. But not all passive solar homes are built this way. Adobe houses, for instance, contain interior as well as exterior mass. For reasons that

FIGURE 1-15
*Brick floors and plaster walls provide thermal mass in this straw bale home in southern New Mexico.*

**TABLE 1-1.**
**Mass Wall Thickness**

| MATERIAL | DENSITY (pounds/cubic feet) |
| --- | --- |
| Concrete | 140 |
| Concrete block | 130 |
| Clay brick | 120 |
| Lightweight concrete block | 110 |
| Adobe | 100 |

will become evident in chapter 5, exterior wall mass is most suitable for hot, dry climates, although it may perform adequately in cold regions if measures are taken to reduce heat migration through the wall into the cold outside environment—for example, by applying rigid insulation to the external surface of the wall or placing insulation in the interior of the wall. If the walls are not insulated, they can lose considerable amounts of heat to the outside and interior comfort levels suffer enormously.

## How Thermal Mass Functions

To design a home properly, you must understand how heat moves and how thermal mass functions.

As noted above, thermal mass absorbs and stores heat. In a solar home, thermal mass acquires heat when it is struck by sunlight. Visible light striking the surface of the mass is absorbed and converted to heat energy. Some of the heat may be radiated into the air around the mass; some may be absorbed by the mass.

Thermal mass also acquires heat from other sources. Warm air flowing over and around a thermal mass wall, for instance, transfers heat to the wall if the wall is cooler than the air. Mass also obtains heat that radiates off other objects in a room. A masonry heater, for instance, radiates heat into a room, which may be absorbed by thermal mass.

Heat absorbed by thermal mass follows a couple different routes. Like any heated object, thermal mass may transfer some of the heat immediately back to the surrounding air. Heat may also migrate inward, toward the cooler interior of the mass. Heat stored in thermal mass remains there until room temperature gently demands its release.

Heat release from thermal mass occurs any time room temperature falls below the surface temperature of the mass. As heat on the surface of the mass begins to flow into the room, heat stored deeper in the mass begins flowing outward toward the cooler surface. When it reaches the surface of the mass, it is released into the cooler room air.

Thermal mass is a simple, highly effective heating system that operates without noisy moving parts that inevitably need repair or maintenance. It relies entirely on the passive flow of heat from warm to cold.

## Distributed vs. Concentrated Thermal Mass

Thermal mass may be incidental or intentional, and internal or external. It can also be concentrated or distributed. Concentrated thermal mass is mass located in one or a few limited areas. Distributed mass is disbursed more or less evenly throughout a house—for example, in the earthen plaster on interior surfaces of the exterior walls of a straw bale house. Concentrated mass tends to service a smaller area. Distributed mass helps maintain uniform thermal comfort. A combination of the two often works best.

## Color and Mass Performance

For many years, solar home designers operated on the premise that thermal mass performs optimally when it is dark colored. In recent years, however, designers have begun to rethink this dogma and now contend that not all mass in a home should be dark colored. In fact, many designs rely on light-colored interior walls and ceilings to help distribute solar radiation more evenly through the interior of a house. Light-colored walls nearest solar glazing, for instance, reflect light onto dark-colored thermal mass located deeper within the structure to ensure greater and more even distribution of heat. (This subject is discussed further in chapter 3.)

## Undermassed and Out of Luck

Because thermal mass stores heat, it helps to reduce fluctuations in the interior temperature of a passively conditioned home. However, if inadequate amounts of thermal mass are installed, heat from the winter sun may cause the daytime temperature to rise into the mid- and upper 80s, making the interior spaces of an "undermassed" passive solar home quite uncomfortable. When the sun sets, heat quickly dissipates out of such houses. The lack of mass to maintain warm interior temperatures therefore results in dramatic temperature swings, often as much as 30°F. Comfort levels decline and back-up heating systems must operate more regularly to maintain comfortable temperatures at night or during cloudy periods.

## Direct vs. Indirect Exposure

Although mass will absorb heat from warm room air, my experience has shown that the most efficient way to capture the sun's energy is to place thermal mass directly in the path of incoming solar radiation. Others agree. Steven Winter Associates, one of the nation's leading solar designers, notes that "storage mass that is heated only indirectly by warm air from the living space requires roughly four times as much area as the same mass in direct sun to provide the same thermal effect." Although placing mass directly in the path of sunlight is desirable, the goal is much easier to articulate than to achieve.

*Floor Mass and Perimeter Insulation.* Floor mass is one of the easiest and most effective ways of placing thermal mass in direct sunlight. A stained concrete slab, a dark tile on a concrete slab, and brick or flagstone on sand all work well, as do adobe floors in natural homes.

When relying on floor mass, be sure to install adequate perimeter insulation—that is, be sure to insulate the foundation walls sufficiently to reduce heat loss. Subslab insulation also helps, as discussed in chapter 2.

*Thermal Storage Walls, Planters, and Interior Partition Walls.* Another highly effective means of providing mass in direct contact with incident solar

*The basic strategy is to design the house so that its own masses—mainly walls and floors—are so placed, proportioned, and surfaced that they will receive and store a large measure of incoming solar energy during the daylight hours and will gently release this stored heat to the house interior during the succeeding night hours or cloudy days.*

PETER VAN DRESSER,
*Passive Solar House Basics*

radiation is the thermal storage wall or Trombe wall, discussed in the introduction. We'll explore the design of Trombe walls in chapter 3.

Thermal mass can also be located in planters and interior partition walls made from masonry materials. If painted dark or plastered with a dark material, they absorb most of the sunlight that strikes them. If they are lighter in color they will reflect sunlight into the home, brightening spaces. Sunlight reflected off a light-colored mass wall can then be absorbed by dark-colored mass located deeper within the home.

## How Much Mass Should a Home Have?

Thermal mass is vital to the performance of a passive solar home. Distributing it correctly is equally important. But how much mass do you need?

As a general rule, the more south-facing glass in a house, the greater the thermal mass requirement. "Although the concept is simple," note the authors of *Guidelines for Home Building*, "in practice, the relationship between the amount of mass is complicated by many factors." Nonetheless, there are some guidelines that can help us determine the right amount of solar glazing in relation to mass, known as the *glass-to-mass* ratio. I'll present them in chapter 3 for each type of solar design.

*Can You Have Too Much Mass?* While you might assume that the more mass you have, the better off you'll be, many passive solar experts think that this isn't necessarily true. The effectiveness of most forms of thermal mass increases proportionately up to four inches. In other words, the thicker the mass the more heat it absorbs, up to a four-inch depth. Beyond that, however, the effectiveness of mass trails off. In fact, a six-inch mass wall is only 8 percent more effective than a four-inch wall. Because thermal mass is generally costly to install, install only what is needed. Going overboard on mass may seem like a good idea, but it's a waste of money experts say. "The most important thing is to have a lot of surface area of mass. Two hundred square feet of two-inch-thick mass is more effective than one hundred square feet of four-inch-thick mass because the diffusivity (heat conduction) and rate of heat transfer from the surface of masonry are relatively low," says Ron Judkoff, director of the Buildings and Thermal Systems Center at the National Renewable Energy Lab in Golden, Colorado.

Not all designers and builders agree with this assessment. Earthship advocates, those who build walls from tires packed with dirt, contend that the thick mass walls of their homes provide massive storage that yields a more stable interior temperature year-round. They are not concerned with supplying heat for two- or three-day periods to compensate for an episodic lack of sunshine. They are designing homes, they say, to provide for much longer periods. Unfortunately, there's not much science to back up either view.

## Water as a Source of Thermal Mass

Water is also a great source of thermal mass. Water holds nearly twice as much heat as standard masonry materials, and even more than less-dense earthen materials such as adobe. However, water gives up heat faster than masonry or earthen materials.

Because water is an effective form of thermal mass, sealed water tubes are sometimes installed in solar houses—often placed in direct contact with incident sunlight. Some manufacturers produce water walls, plastic containers that are installed in walls.

## Thermal Mass and Passive Cooling

Thermal mass plays a key role in passively heated structures. As you will see in chapter 5, thermal mass also plays an important role in passive cooling in many climates. In hot, desert climates, for instance, it serves as a heat sink that absorbs internal heat within a structure during the summer. This heat is purged from the house at night by opening windows. External mass is especially important for maintaining thermal comfort in hot, arid desert regions. Natural building materials such as cob, rammed earth, and adobe provide external mass and therefore represent excellent choices in such environments.

## PRINCIPLE 7: INSULATE WALLS, CEILINGS, FLOORS, FOUNDATIONS, AND WINDOWS

Like many other components, insulation is vital to passive solar heating and cooling. Most solar designers prefer wall insulation ranging from R-22 to R-30 and ceiling insulation of R-40 to R-50, sometimes more (see table 3-1). The lower end of the range is generally recommended for warmer climates and the upper end for colder climates. (R-value is a measure of a materials' resistance to heat movement.)

At a very minimum, insulation should conform to local building codes. In many regions, however, minimum code standards for insulation are inadequate for optimum performance of a passively conditioned home. A far better approach is to design at least to the levels prescribed by the International Energy Conservation Code or, for commercial buildings, ASHRAE 90.2, region-specific recommendations for building envelopes. Most serious solar designers exceed these standards.

Achieving the correct amount of insulation can be difficult. Experts warn against overinsulating a house. Once you've reached optimum levels and taken into account the stability of the insulation, they say, it is time to stop. "Don't go to extremes," note the authors of *Designing Low-Energy Buildings.* "As is the case with any energy-efficient measure, the use of insulation . . . reaches a point of diminishing returns beyond which other strategies become more cost-effective." By using energy-analysis software, such as ENERGY-10 (see

*Water absorbs and stores more heat than masonry materials but dissipates heat more quickly.*

*The thermal resistance of ceilings, roofs, walls, and floors is affected not only by the R-value of the insulation, but also by the resistance of other elements in the construction assembly, framing, sheathing, drywall, etc.*

Guidelines for Home Building

chapter 7), a designer "can compare incremental improvements in shell construction with other measures that may also have merit." Taking into account initial costs and an assortment of other factors important to the project, a designer can make better-informed decisions.

On a personal note, I prefer to superinsulate a house because of research I have seen showing that the commonly used forms of insulation such as cellulose and fiberglass decrease in insulating qualities or R-value over time, primarily as a result of settling or moisture accumulation. Overinsulating compensates for the inevitable reduction in R-value.

Contrary to popular misconception, insulation is as important in hot climates as it is in cold ones. In the southern and the southwestern United States, for instance, summer cooling can be extremely costly—a $300 or higher monthly electrical bill is not uncommon. In such instances, insulation provides great comfort, and, when combined with other passive cooling measures, it can greatly reduce monthly energy bills.

With the cost of natural gas and electricity on the rise, insulation is a wise investment. However, as Mark Freeman aptly points out in his book *The Solar Home*, insulation is only as good as the people installing it and the quality of the job. "Insulation that is crammed in too tightly, has gaps, or becomes wet loses most of its insulating value," he notes.

Some natural building techniques such as straw bale construction provide extraordinarily high levels of wall insulation and are therefore an excellent choice for a passively conditioned home. When two-string straw bales are used and laid flat in the wall, builders can achieve R-values of around 32.

Insulation reduces heat loss through walls in the winter and heat gain during the summer, but other materials and methods are also required to minimize unwanted heat transfers. Special framing techniques and wall construction materials, for instance, reduce bridging loss. Bridging loss refers to the conduction of heat from a building through the framing members in the walls, ceilings, and roofs. It contributes substantially to winter heat loss. Heat can also enter a building this way in the cooling season.

In warm climates, homeowners can reduce summer cooling loads by installing radiant barriers—foil applied to roof decking or roof rafters. Radiant barriers, described in chapter 5, block heat penetrating through the roof of homes on hot summer days. Although the greatest savings occur in summer months, radiant barriers also cut down on heat loss in the winter.

Corner framing techniques can also be modified to reduce heat loss. For a detailed discussion of innovative ways to reduce heat loss and wood use in framed corners, see Charles Bickford's article in the December 1997/January 1998 issue of *Fine Homebuilding*. It is reprinted in *Energy-Efficient Building*, a compilation of articles from *Fine Homebuilding*, published by Taunton Press.

Be sure to insulate foundations, basement walls, and slabs in colder climates as noted earlier. A lot of heat can migrate out of a home through its

Bridging loss refers to heat conduction through framing members in a home, including studs, headers, and all-wood corners. Together, they can form up to 30 percent of the surface area of a wall and therefore represent a large area for heat transfer into a house during the summer or out of a house during the winter.

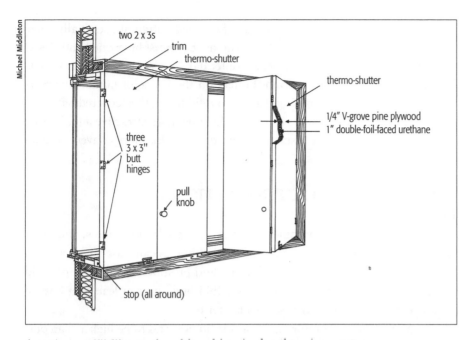

Michael Middleton

two 2 x 3s
trim
thermo-shutter

thermo-shutter

1/4″ V-grove pine plywood
1″ double-foil-faced urethane

three
3 x 3″
butt
hinges

pull
knob

stop (all around)

FIGURE 1-16
*Thermo-shutters made from rigid foam insulation and plywood provide superior insulation for windows in passively conditioned homes.*

foundation. We'll examine this subject in detail in chapter 2.

Window insulation is another important consideration in passive solar homes. Many solar homeowners make the egregious error of leaving south-facing (and other) windows uncovered at night. This is especially common in houses that have huge glass "walls," which are nearly impossible to cover with window shades.

Glass is an extremely poor insulator and therefore conducts an enormous amount of heat to the outside. Because of this, windows can become a major source of heat loss in a highly insulated, air-tight structure, losing up to 50 percent of the heat, according to one estimate. Don't be persuaded to leave your windows uncovered by the fact that you've installed energy-efficient windows. A high-quality double-pane window has an R-value of only about 4. While more efficient than single-pane glass, R-4 glazing still loses tremendous amounts of heat at night, making occupants much more uncomfortable than they need to be and costing you much more than necessary to heat a home.

Cover windows with insulated drapes or shades. Good window coverings can easily add another R-3 to R-4 to windows. Better yet, you may want to consider installing insulated window shades that seal around the edge of the window via magnetic strips. They not only augment the R-value, they help block the flow of cold air from windows and are well worth the expense or time required to make them yourself.

Like so many design features, heat-resistant windows also function during the cooling season, blocking heat gain from warm outside air. Considerable amounts of heat can enter a home this way.

Moving up further on the scale of energy efficiency, some homeowners have installed insulated exterior shutters, mentioned earlier in the chapter.

*Windows can become a major source of heat loss in an highly insulated, air-tight structure, losing up to 50 percent of your heat.*

Although exterior shutters can reduce heat loss, often substantially, they require considerable operator involvement, as they must be opened and closed each day—a loathsome task on a cold winter day. To avoid this problem, some builders install rigid foam thermo-shutters on the inside of homes (figure 1-16). Typically made from plywood glued to rigid foam, then covered with decorative fabric, interior shutters can be made to be quite attractive and, if fitted tightly, are highly effective insulators. And they're a lot more convenient than exterior shutters.

## PRINCIPLE 8: PROTECT INSULATION FROM MOISTURE

Moisture from washing machines, plants, pets, people, aquariums, and other sources in tightly sealed, passively conditioned homes can raise havoc inside a home. It damages window sills and sashes and promotes the growth of mold and mildew on walls and in insulation. Mold build-up can lead to a host of serious health problems, discussed in chapter 6.

In most climates, moisture levels inside homes tend to be higher than outside, due to the ubiquity of internal sources and the fact that interior air is generally warmer than exterior air (warm air holds more moisture than cold air). Like chemicals in our environment and our bodies, moisture tends to move from areas of high concentration to low concentration. As a rule, then, water vapor tends to migrate from the interior of a house to the exterior, passing through ceilings and walls to escape into the outside environment. It enters

FIG. 1-17

*Vapor barriers protect insulation from moisture in all climates. Note that in cold climates, vapor barriers are located toward the inside of the wall (just beneath the drywall). In warm, humid climates, vapor barriers are located toward the outside of the wall, just beneath the siding.*

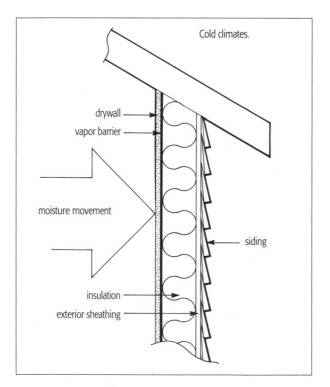

Cold climates.

drywall
vapor barrier
moisture movement
siding
insulation
exterior sheathing

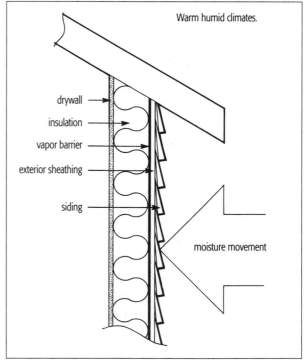

Warm humid climates.

drywall
insulation
vapor barrier
exterior sheathing
siding
moisture movement

walls primarily though cracks around windows, doors, electrical outlets, switches, and other locations.

When water vapor escaping through a wall contacts a cold surface, typically the exterior sheathing of the walls or roof decking, it condenses, dampening, even soaking, the insulation. (The temperature at which water vapor condenses to form liquid water is called the *dew point*.) When wet, most forms of insulation experience dramatic declines in R-value. Even small amounts of moisture can decrease the R-value by 50 percent or more. (The only exceptions are wool, rigid foam, rock wool, and plastic bubble insulation.)

Besides greatly decreasing the R-value of the insulation, moisture escaping through ceilings and walls can collect on roof decking and wall sheathing, causing it to warp and rot. An entire roof may need to be replaced as a result. Furthermore, moisture collecting in ceiling insulation can drip down on drywall, causing considerable damage.

Fortunately, there are several simple, cost-effective ways to reduce the flow of moisture through walls and ceilings. One of the most effective is the installation of a vapor barrier (figure 1-17). Further details are presented in chapter 2.

## PRINCIPLE 9: SEAL HOUSES AGAINST AIR INFILTRATION BUT PROVIDE ADEQUATE AIR EXCHANGE

Unbeknownst to most people, many of our homes are a virtual Swiss cheese. According to one source, there are so many air leaks in an average American home that added up they would be equivalent to an open window measuring 3 feet by 3 feet. This leakage around windows, doors, and foundations wastes enormous amounts of energy and can make our homes very expensive to heat and to cool and extremely uncomfortable.

"Sealing a house carefully to reduce air infiltration—air leakage—is as necessary to energy conservation as adding insulation," according to the authors of *Guidelines for Home Building*. "Even a small opening can allow heat to bypass the insulation and lead to big energy losses," they add. Sealing a home well reduces energy demand during both the heating and the cooling season, lowering utility bills and increasing comfort levels.

Expanding foam insulation and caulk should be used to seal all penetrations in the walls or foundations created by electrical wires and plumbing. Care should also be taken to seal around doors and windows and along the top plate and sill plate. A house wrap, plastic applied to the exterior sheathing, helps reduce air infiltration, as do vapor barriers. Both are discussed in more detail in chapter 2.

It is also important to seal ducts in back-up heating and cooling systems. Leaky ductwork can result in the release of substantial amounts of warm or cold air into unconditioned spaces, thus wasting a considerable amount of money and fuel. Studies performed in Florida, in fact, indicate leaky ducts of

*Sealing a house carefully to reduce air infiltration–air leakage– is as necessary to energy conservation as adding insulation.*

Guidelines for Home Building,
Sustainable Buildings
Industry Council

FIGURE 1-18

*In rectangular floor plans, less frequently used rooms are typically placed on the north side of a house. Rooms used more frequently are placed on the south side for optimal comfort.*

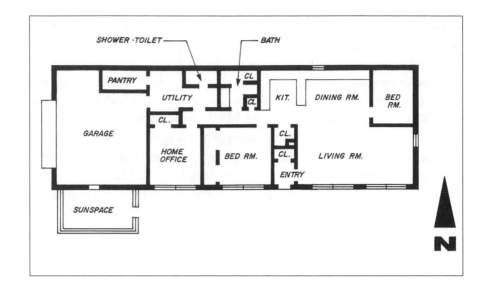

**SUNGLASSES AND BASEBALL CAPS ON CHRISTMAS**

An architect I met at a conference designed a passive solar home for her family. With a two-story south-facing glass wall, this elegant home offered tremendous views and a wonderfully open feeling. It per-formed admirably during the winter, providing plenty of heat. However, solar gain came with its price. For Christmas the first year they lived in the house, her elderly parents came to visit. On Christmas morning the family gathered around the Christ-mas tree to open presents. With blinding sunlight streaming into the living room, her parents were forced to wear sunglasses and baseball caps to shield their eyes while the family exchanged gifts.

heating and air conditioning systems accounts for up to 25 percent of a house's annual heating and cooling load. Many green builders take time to seal ducts. For best results, they recommend the use of water-based mastics, rather than duct tape. Although they cost more, they last much longer.

Because sealing a building envelope and duct work can be tricky, many contractors hire specialists to do the job on the theory that they can provide a higher-quality job. Remember, however, that increasing the tightness of a house beyond the recommended level may improve energy performance, but it can also create serious indoor air-quality problems. In energy-efficient pas-sively conditioned homes, proper air exchange required to protect indoor air quality is often achieved by sealing a house, then installing a ventilation sys-tem. We'll explore this topic in chapter 6.

## PRINCIPLE 10: DESIGN HOMES SO THAT EACH ROOM IS HEATED DIRECTLY OR IS ACCESSIBLE TO SOLAR HEAT

The ideal solar design for most climates is one room deep, so that incoming sunlight can extend well into the interior of each room of a home. In essence, each room becomes its own self-contained solar unit. In addition to providing an encapsulating warmth, this arrangement eliminates the need to collect heat in one location and transport it to another unheated or less heated space.

Unfortunately, this strategy requires a linear house design, which is not always conducive to efficient internal traffic flow. Long houses may also be less-than-appealing visually and may not be appropriate for smaller lots in cities and towns.

To address these problems, passive solar designers often turn to rectangu-lar or square home designs. In such instances, rooms that are frequently occu-pied—and thus require warmer temperatures, such as living rooms, bedrooms, and home offices—are typically located along the south side of the house. Spaces that require less heat, including corridors, utility rooms, laundry rooms, pantries, and workshops, are situated along the north side (figure 1-18). These spaces receive heat primarily by convection from the sun-warmed rooms along the south side of the building.

## Open Floor Plans

Open floor plans, popularized by Frank Lloyd Wright, can help distribute heat throughout a passive solar home via convection. Open floor plans offer many additional benefits beyond natural heat distribution. As architect Sara Susanka points out in her books *The Not So Big House* and *Creating the Not So Big House,* open floor plans can make smaller spaces appear larger than they really are. Thus, environmentally conscious homeowners can build smaller passive solar homes without creating a cramped feeling. Building smaller dwellings reduces the resources that must be marshaled to create and maintain shelter. Smaller homes mean less clear cutting, less mining, less paint, and less fossil fuel energy for making building materials and maintaining comfort. Smaller size also slashes construction costs and reduces the amount of land that must be disturbed to create human shelter.

Although smaller homes with open floor plans offer substantial benefits to people and the planet, there are some drawbacks to this approach. Open floor plans, for example, make it difficult to zone heat a house. On cold, cloudy days when solar gain vanishes, more energy will need to be expended to provide heat to active-use zones.

Noise is another problem. Open designs can be much noisier than more conventional floor plans. Children playing in the family room may disturb quiet conversations or study in adjoining rooms. When designing a home with

**FIGURE 1-19**
*The window shades are drawn on this passive solar home during the winter to prevent sun drenching and overheating, an all-too-common but avoidable problem. By careful design of homes, architects and designers can permit solar gain without rendering living space useless during daylight hours.*

*One of the most overlooked aspects of solar design is the sun-free zone.*

an open floor plan, be sure to distribute space within the structure to mitigate noise or include rooms that can be closed off to intrusive sounds.

## PRINCIPLE 11: CREATE SUN-FREE SPACES

When designing a passive solar home, it is imperative that we pay as much attention to creating sun-free zones as we do to creating sun-bathed interiors for solar gain. Although this design principle may seem contradictory—flying squarely in the face of some of the basic solar design principles outlined above—it is vital to ensuring comfort and creating a home with more usable space.

In many passive solar homes built in the 1970s and 1980s in the United States, living spaces were often drenched in sunlight during much of the heating season. Although these designs provided plenty of heat, rooms were so brightly illuminated that they often became unusable during daylight hours. Many a disappointed homeowner turned to heavy window shades to block the sun (figure 1-19). While this tactic makes a room usable, it dramatically cuts down on solar gain.

*Sun drenching* is my term for this widespread and sometimes severe problem found in old and some new passive solar homes. Excess sunlight streaming into rooms can produce glare on televisions and computer screens, resulting in severe eyestrain and headaches. In some cases, glare makes screen viewing impossible. Sunlight can bleach carpets and furniture, too.

### Solving the Problem of Sun Drenching

What is a person to do? Sun-drenched rooms are the main goal of passive solar, aren't they?

Fortunately, some simple design strategies can be implemented to ensure solar gain while producing sun-sheltered space within a solar house. The most effective are design approaches: alterations of the floor plan that create sun-

FIGURES 1-20
*This floor plan of the author's home shows various design features included to reduce sun drenching and create useful living space. Note on the garage side how the front airlock (an attached sunspace for solar gain) shields the living room. A wide hallway down the south side of the house protects the kitchen and living room. Partition walls separating the bedrooms from the hallway shield bedrooms, and the bathroom shields the author's office.*

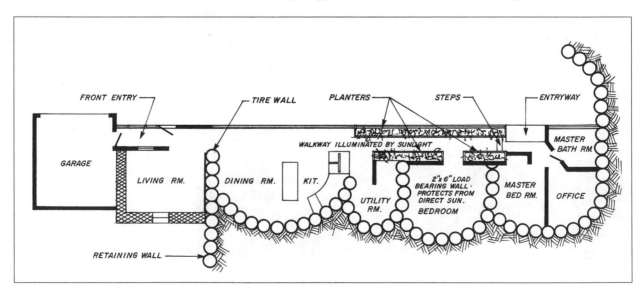

FIGURES 1-21
*This airlock serves many functions. It gathers solar energy that is released into nearby rooms, protects the living room from excess sunlight, reduces cold air entering the house when visitors and family members enter and leave, and serves as a repository for jackets, shoes, and boots.*

free zones. Planters, hallways, partition walls, entryways, and other design features may all be used toward this end. In my house, the front airlock shields the living room from direct sun, greatly reducing glare on the TV (figures 1-20 and 1-21). My office is sheltered by a bathroom located adjacent to it along the south wall. The bedrooms are protected by a growing area with planters and a divider wall.

Sunlight can also be directed into interior stairwells where bright light meets little, if any, objection. Sunlight can be directed onto masonry heaters, as well (chapter 4). Trombe walls are very effective in reducing sun drenching, as are clerestory windows, which can be used to deliver sunlight to the back walls of a room. Because it is so important, we'll visit this subject in more detail in chapter 3.

## PRINCIPLE 12: PROVIDE EFFICIENT, PROPERLY SIZED, ENVIRONMENTALLY RESPONSIBLE BACK-UP HEATING

Solar designs typically require some form of back-up heat. Even if a house is designed to provide all of its heat from the sun, building departments will very likely require the installation of some sort of back-up heating system. Because passively conditioned houses are well insulated, free of leaks, thermally stable, and heated primarily by solar energy, the heating system won't need to be very large.

Whatever you do, don't let a heating contractor unfamiliar with the full potential of passive solar design decide the size of the heating system you should install. They are generally inclined to oversize back-up heating systems to provide a measure of safety and protect them from potentially angry customers who call to complain about an inadequate heating system.

*Owners report great satisfaction with the performance of their passive solar homes and rate comfort and quality of their living environment equally high. These energy-efficient homes are bright, sunny, quiet (no operating noise), comfortable, and affordable.*

The Passive Solar Industries Council

FIGURE 1-22
*Earth berming, as along the north side of this solar home, protects homes from excessive heat loss during winter months and heat gain during the summer months. It therefore helps to maintain even temperatures year-round.*

*Earth sheltering refers to partial (earth bermed) to complete burying (underground) of a house. Most underground homes present an exposed south face (in the Northern Hemisphere) to permit sunlight to enter and to permit a view of the outside world.*

Oversizing a system adds unnecessarily to initial costs. In addition, larger systems cycle on and off more frequently than smaller systems, burn more fuel, and thus cost more to operate. The customer ends up paying a lot more to provide the heating contractor with peace of mind.

Choices of back-up heating systems are many and varied. We'll look at them in chapter 4 and explore criteria for selecting the most environmentally sound and healthful options.

## PRINCIPLE 13: PROTECT HOMES FROM WINDS BY LANDSCAPING OR EARTH SHELTERING

The south-facing windows of a home gather valuable sunlight to heat a home and keep its occupants warm and comfortable, even in the worst of weather. But comfort also requires heat-conserving measures. Insulation, shades, and thermal mass, for example, all help to retain heat and boost performance of a solar house. But other strategies can be of service as well. Earth sheltering, for instance, protects a home from heat loss.

Earth sheltering involves partial to nearly complete "burial" of a house. Partial sheltering, also known as *earth berming,* is achieved by burying the lower parts of the exterior walls of a house, creating a berm typically three to four feet high. This is a valuable means of protecting a house situated on flat terrain (figure 1-22).

Houses can also be built more fully underground—that is, against a hillside with dirt pushed up onto a specially built roof (figure 1-23). The south side is left open. Although the thought of living "underground" may seem depressing, well-designed underground homes are far from dark caves. With proper design, a home such as this can provide plenty of light and generous views of the outdoor world along the exposed solar face of the building. Most people who tour my house are shocked when we reach the office and bedrooms, which are "completely" underground (the walls and roof are covered with dirt; the south face is exposed). They're amazed at the brightness of the rooms.

By nestling the north side of a house into a hillside, perhaps burying the roof as well, a homeowner can prevent winter winds from stripping heat away from a structure. Dave Johnson, who built my house with me, told visitors to think of a house as a person wrapped in a blanket facing the sun. The earthen blanket protects against the wind, while being open slightly in front to let the heat in.

Earth sheltering allows us to benefit from thermal energy stored in the Earth. Below the frost line, the ground stays a fairly constant 50°F, plus or minus a few degrees, depending on where you live. An earth-sheltered house takes advantage of this phenomenon and remains at a fairly constant temperature year-round. In the winter, earth-sheltered houses require very little additional heat to raise interior temperatures into the comfort zone. Solar energy can provide most, if not all, of the necessary heat to boost temperatures by 20 degrees F. Compare this to a stick-frame house that is exposed to the elements. If unheated, the temperature inside such a house falls pretty close to the ambient (outdoor) temperature. Heating it to the comfort zone is much more difficult than heating a home that is tucked inside a 50 degree F earthen blanket.

Earth sheltering tends to keep a house cooler in the summer, too. It may be 100 degrees F outside, but the house stays a comfortable 70°F thanks to its cool earthen wrap. In the summer, the temperature inside a wood-frame house can bake you alive.

In his book *The Earth-Sheltered House*, Malcolm Wells sums up the virtues of earth-sheltered architecture. "It works," he says. "It really works.

FIGURE 1-23
*Far more effective than earth berming is full earth sheltering. This home, the author's, is built into a hillside with dirt pushed up onto the specially designed, waterproof roof. Elk routinely graze on the living roof. Insolation and dirt help to deaden external sounds and maintain even temperatures year-round.*

And, in addition to having a green 'footprint,' every *well-designed* underground (earth-sheltered) building is also silent, bright, dry, sunny, long-lasting, easy to maintain, easy to heat and cool, and fire safe" (italics mine).

Earth-sheltering is not without problems—largely resulting from improper design or construction (or both). If improperly designed and built, roofs can leak or, worse yet, cave in. Water can accumulate in the backfill around a house and can seep through the walls, if appropriate precautions are not taken. Backfill around the house can settle. As it sinks, it may strip rigid insulation board or waterproofing off the wall and can rupture buried pipes.

The moral of the story, says Malcolm Wells, is not that earth sheltering should be avoided, but that it should be done right. If this idea interests you,

## PASSIVE SOLAR HEATING AND COOLING: DOES IT MAKE ECONOMIC SENSE?

Passive solar heating and passive cooling can save homeowners hundreds of dollars a year by reducing fuel bills. But is the cost worth it?

According to Ron Judkoff, director of the Buildings and Thermal Systems Center at the National Renewable Energy Lab, passive solar can increase the cost of building a new home by from 0 to about 3 percent. On a $200,000 home, then, passive solar heating could boost the cost by as much as $6,000. That's the high end, however. As Judkoff quickly points out, because many building codes now call for much more energy-efficient windows, walls, ceilings, and foundations than in the past, passive solar frequently adds very little or nothing to the cost of a new home with careful holistic design. And the economic gains can be quite impressive.

Judkoff bases his cost estimates and savings in part on a series of

case studies of passive solar homes sponsored by the National Renewable Energy Lab and the American Solar Energy Society. Data was collected on passive solar homes from a diversity of locations— Arizona to Wisconsin, Indiana to Maine, and Massachusetts to North Carolina. In this study, the results of which are summarized in the accompanying table, they found that the additional cost of building a passive solar home ranged from zero dollars to $6,000, while the savings from passive solar heating fell within a range of $220 to $2,255 per year. Over a thirty-year period, the total savings *at current energy costs* reaped the owners of these homes $7,000 to $67,000.

But energy prices are not likely to remain stationary. In fact, major increases in natural gas and oil prices are very likely. Although no one knows how high oil and natural gas prices will climb in the

coming decades, energy expert Randy Udall hesitantly offered a conservative estimate of 60 to 70 percent over the next twenty years.

Personally, I suspect that peaks in global oil production and U.S. natural gas production will send fossil fuel prices through the roof over the next twenty to thirty years. After all, the cost of natural gas in Colorado doubled in 2001 alone as prices across the nation skyrocketed. (Prices went down 40 percent the following year.)

Those who invest in passive solar will be amply rewarded. Using the data from NREL and ASES case studies and assuming a 5 percent per year annual price increase in natural gas, annual savings at the low end, $220 per year, would increase to over $900 per year by 2032. At the high end, the current annual savings of $2,255 would increase to over $8,800 by 2032. Total savings over the thirty-year period at the low end would be an

exorcise all images of living in a cave and get a copy of *The Earth-Sheltered House* or *The Complete Book of Underground Houses* by Rob Roy. I'll talk more about earth sheltering in chapter 3 and chapter 6.

Another effective way to protect a house from wind, if earth sheltering isn't an option, is to situate your home in the protection of a grove of trees or some natural windbreak, such as a small hill. You can also plant your own windbreak. For more on this subject read chapter 15 of my book, *The Natural House*. Anne Simon Moffat and Marc Schiler's *Energy-Efficient and Environmental Landscaping* provides even greater detail.

astounding $14,600. At the high end, the thirty-year savings would be $141,400. Quite a return on investment.

If energy prices increase 10 percent per year, the $220 per year savings would increase to nearly $3,500 per year, reaping a homeowner a thirty-year net savings of $36,200. The $2,255 per year savings would jump to nearly $40,000 per year with a total thirty-year savings of nearly $404,000.

Although some energy analysts predict much lower increases, the fact is no one knows for sure what will happen to fuel prices. This we do know, however: Shortfalls are likely to wreak havoc on future fuel prices. If this is true, millions could benefit economically from including passive solar in new and existing homes. It may be one of the very few—and surely one of the most economical—choices we have for maintaining indoor comfort in the coming decades and minimizing the effects of instability in fossil fuel prices.

**TABLE 1-2. Energy Costs and Savings Resulting from Passive Solar Design**

| Location (date construction was completed) | Additional cost | Reduction in heating and cooling cost | Annual savings in fuel bill | 30-year savings based on current fuel costs | 30-year savings based on projected 5% and 10% increase in fuel costs per year |
|---|---|---|---|---|---|
| Jonesport, Maine (1988) | $1,000 | 70% | $300 | $9,000 | $19,900 - $48,700 |
| Falmouth, Massachusetts (1995) | $3,500 | 82% | $1,260 with PV system for electricity | $37,800 | $76,894 - $207,750 |
| Burlington, North Carolina (1990) | $5,000 | 64% | $840 | $25,200 | $55,339 - $138,160 |
| Naperville, Illinois (1984) | $3,000 | 72% | $550 | $16,500 | $36,540 - $90,430 |
| Stevens Point, Wisconsin (1995) | $6,000 | 70% | $600 | $18,000 | $42,179 - 98,690 |
| Hanover, New Hampshire (1994) | 0 | 95% | $2,255 | $67,650 | $141,390 - $403,976 |
| Andover, Connecticut (1981) | 0 | 58% | $958 | $28,740 | $63,650 - $157,675 |
| Santa Fe, New Mexico (1985) | 0 | 81% | $220 | $6,600 | $14,614 - $36,196 |

FIGURE 1-24

*A modern passive solar home.*

FIGURE 1-24
*A modern passive solar home.*

## PRINCIPLES OF PASSIVE SOLAR HEATING AND COOLING

1. Choose a site with good solar exposure.
2. Orient the east-west (long) axis of the home within 10° of true south.
3. Concentrate windows on the south side of the house.
4. Minimize east and west glazing.
5. Provide overhangs and shade to regulate solar gain.
6. Provide a sufficient amount of well-positioned thermal mass for heating and cooling.
7. Insulate ceilings, walls, floors, and foundations.
8. Protect insulation from moisture.
9. Design house so that rooms are heated directly and for optimal natural heat distribution.
10. Create sun-free zones for computer work and television-watching.
11. Seal leaks and cracks to reduce air infiltration but ensure adequate ventilation for fresh air.
12. Provide efficient, properly sized, environmentally responsible back-up heating and cooling systems.
13. Protect homes from winter winds by landscaping, earth sheltering, and other measures.
14. Design interior space to correspond with solar gain and living patterns.

## PRINCIPLE 14: SYNCHRONIZE DAILY LIVING PATTERNS WITH SOLAR PATTERNS

One of the components of an intelligent, integrated passive solar house is a design that seeks to synchronize living patterns with solar patterns. Consider an example. Suppose you live in a two-story rammed-earth home. You are one of those creatures who likes to rise early, perhaps with the sun, and like many of us you want to wake up to a warm room. By placing your bedroom on the southeast corner of the second story, you can take advantage of early morning sunlight. If you then shower and head to the kitchen for breakfast, place the kitchen on the first floor in the southeast corner, so that it is warm by the time you arrive.

Designing the interior space to correspond with daily activities and the sun's predictable path across the sky not only provides comfort, it saves money and resources, as it reduces your need for back-up heat and electricity for light. The sun serves you as you move through your day. Combined with the other principles of design (summarized in the sidebar) this simple measure can help designers produce passively conditioned homes that provide a lifetime of comfort with little, if any, outside energy.

## THE MANY BENEFITS OF PASSIVE CONDITIONING

Passive heating and cooling can provide clean and reliable comfort year-round. Although the strategies described here may add to the initial cost of a home, passive solar heating and cooling (if done correctly) often save considerable amounts of money over the long term. Moreover, passive heating and cooling can serve as a hedge against inflation resulting from escalating fuel bills. As fossil fuel prices inevitably rise, those who avail themselves of natural conditioning will surely have a leg up on those who persist in energy-intensive comfort.

Another benefit of passive conditioning is that it provides a measure of independence that appeals to many people the world over. When an ice storm

**FIGURE 1-25**
*These homes illustrate a small portion of the range of architectural styles amenable to passive solar design.*

sends the electric lines crashing down, furnace fans in nonsolar homes cease to run. It doesn't matter to a solar homeowner as a passively heated home requires no fans. Occupants will stay cozy despite the subzero temperatures while their neighbors suffer for their dependence on the costly and highly vulnerable centralized electric power grid.

Passive conditioning provides other benefits as well. It can, for instance, substantially increase the marketability of a home. As energy prices rise, a home that provides comfort at a substantial monetary savings is likely to enjoy a noticeable market advantage over conventional homes strapped with high monthly utility bills. As the world's fossil-fuel energy outlook dims, the prospects for energy-efficient passive solar homes only brighten.

The superior aesthetics of passively heated and cooled homes can likewise contribute to their marketability. While many buyers are not impressed by high levels of insulation in the building envelope or high-efficiency back-up heating systems, a sunny, open design and brightly illuminated living areas heated by south-facing windows are usually a strong selling point. As the authors of *Passive Solar Design Strategies* point out, "Windows in general are popular with home buyers." Successful passive solar designs can transform windows into energy producers instead of energy liabilities. In addition, according to the Sustainable Buildings Industry Council (SBIC) "High-efficiency heating equipment can account for significant energy savings—but it won't be as much fun on a winter morning as breakfast in a bright, attractive sunspace." Bottom line: The more obvious design elements of passively heated and cooled homes, such as windows and open floor plans, are not only functional, they are highly appealing. Some builders, in fact, make no reference to "passive solar" when advertising the homes they've built, preferring instead to present their stylish homes as examples of state-of-the-art energy efficiency.

Another advantage of passive solar is that it is compatible with many architectural styles. When many people hear about solar they imagine contemporary solar homes, such as the one shown in figure 1-24. However, even log cabins and Victorian-style homes can be designed to effectively utilize free solar energy (figure 1-25). The opportunities for integrating passive solar heating and passive cooling in buildings are unlimited. The benefits last a lifetime and will help make this world a better place for all who live here.

# ENERGY-EFFICIENT DESIGN AND CONSTRUCTION

IN THE LATE 1970S, I purchased my first home. I was an assistant professor at the University of Colorado in Denver, teaching biology and environmental science. I knew a little bit about solar energy, and was ready to make the shift from apartment living to home ownership. The house I purchased was an attractive bungalow built in 1925. It wasn't a passive solar structure, but it had a good southern exposure.

Soon after moving in, I started to work on the house. I began by installing solar panels for domestic hot water. I then removed some rather large, leaky north-facing windows, beefed up the attic insulation, and sealed numerous air leaks. Next, I built a small attached sunspace on the south side of the house.

This was my first attempt at passive solar heating. I based the design not on science and solar engineering, but on pure guesswork. Not surprisingly, the attached sunspace performed poorly. It was not large enough to have a noticeable effect on room temperature. Had I known more, I would have constructed a space commensurate with the heating requirements of the house and designed a system to move the air in and out of the sunspace more effectively than the portable fan I mounted in the kitchen window that opened into the sunspace.

Many other individuals like myself experimented with passive solar heating in the 1970s and 1980s. Venturing boldly into the field, designing with inadequate information and inspired by the apparent simplicity of passive solar, we learned that passive solar heating is much more complicated than many of us initially thought.

Many early passive solar houses were built with too much south-facing glazing and too little mass. As a result many of them overheated. "Some of these homes overheated to the point that it damaged the glazing components, resulting in leaky windows," notes one solar energy expert who asked to remain anonymous. "People ended up using interior shading in an effort to keep sunlight and heat out, which isn't all that effective. Or they opened windows to vent the heat and admit cool outside air, and the heat that was supposed to carry them through the night was lost. As a result, the furnace had to run more, eliminating most of the cost savings." When asked what the central problem of early passive solar designs was, my colleague responded, "I think many homes were designed on the principle that if a little glass is good, a lot must be better." He went on to say, "I believe the root causes of the early failures was the well-intentioned but misguided goal of trying to design for maximum solar heating, at or near 100 percent, which is a tough goal to achieve in a passive solar home." Another reason many homes failed to live up to expectations is that they contained inadequate insulation.

This book aims to help others avoid the mistakes of the past. This chapter looks at one of the most important aspects of passive design, energy efficiency. Household energy efficiency, the use of energy in our homes in ways that produce maximum comfort with minimum waste, is a key to successful passive solar heating and cooling. As you will see, passive heating and cooling are complementary: Most energy-efficiency measures that contribute to one goal also contribute to the other. Wall and ceiling insulation are obvious examples. Many other measures also pare down the energy required to heat and cool a home. Some, such as highly efficient lighting or energy-miserly appliances, may seem unrelated to heating and cooling, but the vast majority of the measures designed to make a home more efficient in the use of electricity and other fuels play a part in achieving greater comfort. Efficient lighting, for example, produces less heat than standard lighting strategies and thus helps reduce cooling loads.

By reducing heating and cooling loads, energy-efficiency measures result in two significant dividends: They make passive heating and cooling easier to achieve, and they reduce or eliminate the need for back-up heating and cooling. Energy efficiency is not only vital to achieving comfort in a passive design, it is one of the main reasons why passive conditioning works in most instances.

Before we delve into the specifics of energy-efficient design and construction, it is important to note that energy-efficient design and construction are easier these days thanks to the widespread adoption of more stringent energy codes. Because of this, energy efficiency is now standard practice on many areas. Furthermore, because many efficiency measures are required by code, it won't cost much more to build a passively heated/passively cooled home. In some cases, costs are identical. Whether it costs more or not, it is important to remember that passive conditioned homes save money and improve comfort levels immediately and continue to bestow these benefits every day and night over the lifetime of a building.

In this chapter, we will concentrate on energy-efficiency measures that are familiar to many homeowners, builders, and designers but not well understood. I will also present information about new materials and methods that expand your understanding of energy-efficient design, including energy-efficient foundations, high-performance windows, and insulation. My goal is to provide background material that will enable you to make informed choices as you design and build passively conditioned homes. After you have read this material, you may want to obtain a copy of one of Joe Lstiburek and Betsy Pettit's *Builder's Guides* for specifics and building in either cold, humid, or arid climates. They are listed in the Resource Guide.

## FOUNDATIONS FOR PASSIVELY CONDITIONED HOMES

Foundations are extremely important to all homes, as they provide a solid base on which a structure sits, resisting movement and keeping the structure away from moisture on and in the ground. Unfortunately, very little attention is paid to the energy efficiency of foundations. As you will soon see, foundations can serve as an avenue for heat loss and can undermine the success of a passively heated home. They can also transfer heat into a building during the cooling season, although this is a lesser concern than wintertime heat losses.

In this chapter, I will examine some of the more common foundation types and describe the many ways they can be built more efficiently. I'll also examine some new developments, such as insulating concrete forms and frost-protected shallow foundations, that offer superior insulation.

When selecting a foundation, the designer, builder, and homeowner must consider many factors, including (1) structural stability, (2) energy efficiency, (3) source and impact of the raw materials, (4) labor, and (5) costs.

### Concrete Foundations

Many architects and contractors build their homes on slab-on-grade foundations or the standard concrete stem wall and footer. The slab-on-grade foundation is a foundation and slab poured together, and is also known as a *monolithic pour* (figure 2-1a). The thickened edge of the slab supports the weight of the walls and other overlying structures including the roof and second-story floors. The slab forms the subfloor and is typically covered with tile, carpet, or wood flooring, although the concrete can be stamped or stained. In solar homes, the slab also doubles as thermal mass.

The standard concrete stem wall and footing shown in figure 2-1b is a simpler affair. It consists of a flat wide bottom section, the footing, that distributes load over a larger area (much like our feet). The stem wall is a narrower poured concrete wall that transfers

FIGURE 2-1
*Foundations come in many varieties. The (a) slab-on-grade foundation and (b) standard concrete foundation with stem wall and footing are the most commonly used in conventional construction.*

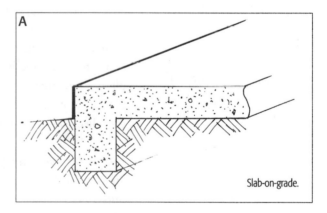

A

Slab-on-grade.

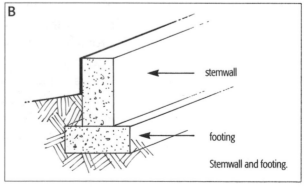

B

stemwall

footing

Stemwall and footing.

weight to the footing (much like our legs). Like the slab-on-grade foundation, the stem wall and footing foundation are reinforced with steel rebar.

If you are considering a concrete foundation, you can take specific steps to enhance its environmental performance. The first has to do with the materials you use, and the second has to do with enhancing their energy performance.

*Fly-Ash Concrete.* One way to improve on concrete foundations is to use concrete manufactured from fly ash. Fly ash is a waste product from coal-fired power plants. Captured by pollution control devices, fly ash is produced in abundance in the more developed countries and is typically dumped in landfills, where toxic compounds may leach into groundwater.

Rather than discard this potentially harmful material, fly ash can be substituted for 15 to 30 percent of the Portland cement in concrete, producing fly-ash concrete.

Fly-ash concrete offers two main benefits. It reduces the embodied energy of concrete—decreasing the total energy required to build a house—and it puts a waste product to good use, preventing its burial in landfills. Fly-ash concrete is also smoother, denser, more workable, and less permeable than conventional concrete. Because of these benefits the U.S. government recommends the use of fly-ash concrete in all federally funded projects. Fly-ash concrete is also widely available.

Despite its benefits, some builders are concerned about potential health effects. They note, rightly so, that fly ash contains an assortment of toxic materials, mostly heavy metals such as lead and mercury. Entombed in concrete, however, these metals probably pose little threat to homeowners. If fly-ash concrete affects the health of anyone, it would most likely be that of the people who manufacture the material and those who pour and finish concrete. (There's not much evidence to support either view.) Do your research when you check out this product.

*Enhancing the Energy Efficiency of a Concrete Foundation.* Slab-on-grade and standard stem wall foundations leak substantial amounts of energy to the outside environment, especially in colder climates. Their inefficiency results from the fact that concrete is a poor insulator and an excellent conductor of heat and that foundations are in direct contact with the cold ground. Heat streams out of foundations in the winter, moving by conduction to the colder external environment.

Insulating concrete foundations makes them more energy efficient not only in the heating season, but also in the cooling season. Because methods of insulating concrete foundations are identical to those used to insulate foundations made from concrete blocks, the next topic, we'll explore foundation insulation in detail after we look at concrete blocks.

*Although concrete foundations are reliable and effective, even in earthquake zones (with proper steel reinforcement), they require enormous amounts of concrete, a material with fairly high embodied energy.*

## Concrete Blocks and More Environmentally Benign Alternatives

Standard concrete blocks are also used to build many foundations. Blocks are laid on well-compacted subsoil, reinforced with rebar and mortared in place with successive courses laid in a running bond for maximum strength.

Like concrete, concrete blocks can be made from fly-ash concrete. If you are interested in this product, even to build a planter or a mass wall, contact local suppliers to see if any of them sell blocks manufactured from fly-ash concrete.

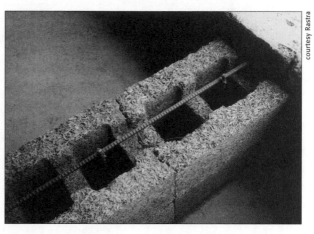

FIGURE 2-2
*Faswall blocks can be used to build foundations and walls. These lightweight blocks are made from cement and wood particles, a waste product. They are relatively easy to work with, but must be insulated to reduce heat loss.*

Like standard concrete foundations, concrete block foundations offer little, if any, insulation. Heat is readily conducted outward through the blocks during the winter and inward during the summer. To thwart this predictable heat flow, some builders fill the cores of concrete blocks with high-pressure foam, polystyrene beads, and vermiculite, or install specially designed foam inserts that fit tightly inside the blocks to reduce heat flow. Although high-pressure foam seems to work best, computer simulations and field studies show significant heat flows continue through the mortar joints. This is a form of bridging loss. To reduce bridging loss, it is best to build a block wall, then apply external foam insulation to the foundation wall, a topic discussed shortly.

A better option than a standard concrete block from an environmental perspective is the Faswall block (figure 2-2). Used in Europe for more than sixty years, Faswall blocks consist of concrete and wood particles (waste products) molded into blocks. They are used to construct foundations, exterior and interior walls, and other structures. They are lightweight and thus relatively easy to work with. As with ordinary concrete blocks, insulation is required to reduce heat loss.

Yet another product worth considering is the Rastra block. Made from cement and foam made of recycled plastic, Rastra blocks can be used to build foundations and walls. Like Faswall blocks, Rastra blocks are lightweight and easy to handle. Moreover, they cut easily with a saw. Both horizontal and vertical rebar may be placed in the blocks for structural stability, as with any block. Stucco adheres well to the porous surface of the Rastra block. Like the Faswall block, insulation will be required to safeguard against heat loss.

## Insulating Concrete or Concrete Block Foundations

When building a passive solar home, foundation insulation is required in all but the warmest climates. Although insulation adds to the cost of a foundation, its benefits far outweigh any additional capital outlay, as an uninsulated foundation serves as a major conduit for heat escape. Insulation greatly

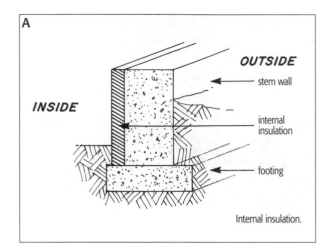

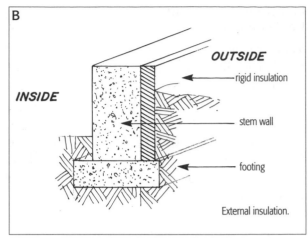

FIGURE 2-3
*Rigid foam can be applied on interior or exterior surfaces of foundations.*

curtails the "energy nosebleed," making a home more comfortable and easier to heat. The colder the climate, the more insulation is required.

Stem walls and slab-on-grade foundations are typically insulated with rigid foam, which is applied externally or internally, as shown in figure 2-3. In exterior applications, the foam is laid against a waterproofed stem wall or slab, extending about two to four feet below the top of the foundation wall. Dirt is then backfilled to hold the foam in place. In this position, rigid foam retards heat flow and also helps waterproof the foundation. Rigid foam can also be applied directly to the inside surface of stems walls.

Basement walls are also insulated internally and externally. In cold climates, external rigid foam insulation should extend at least four feet below grade, and should be about two to four inches thick. The colder the climate, the thicker the insulation. Below the four-foot level, you can reduce the thickness of the insulation by half. (Be sure to investigate the local energy codes to determine specifications for your area.)

Basements can also be insulated internally using rigid foam or bubble insulation, such as Reflectix, a relatively new product consisting of plastic bubbles sandwiched between reflective layers of aluminum. These products are typically installed between furring strips attached to the interior surface of basement walls. Drywall or paneling is then attached to finish the wall surface.

### External vs. Internal Insulation of Foundation Walls
Of the two options, external insulation is generally the easiest and cheapest option in new construction. As in many aspects of building, however, the "easiest and cheapest option" does not always translate into "the best one." Here's why in this instance: External insulation generally does not perform as well thermally as internal insulation. Unless care is taken, external insulation strategies tend to leave pathways to the outside environment, allowing heat to escape. Comparing heat flows in figure 2-4a (external insulation) with figure

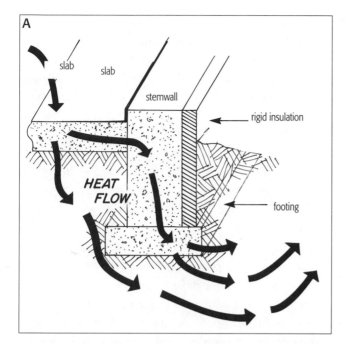

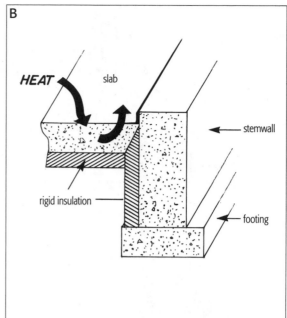

2-4b (internal insulation) illustrates the differences in heat loss between these options.

External insulation may also be subject to damage. Termites and carpenter ants may bore into some forms of rigid foam. Although foam is not a source of nutrition for them, it does provide access to the rest of the house, much of which is quite edible. Insect burrows reduce the structural integrity of the insulation and the R-value as well.

To prevent insect damage, some foam insulation manufacturers impregnate their products with chemical repellent, boric acid being the most common. AFM Corporation in San Diego, a leader in the manufacture of healthy building materials, has been adding borate repellents to some of its rigid foam insulation products since 1990. Unfortunately, when exposed to ground moisture chemical repellents may slowly leach out of the foam, rendering the protection ineffective.

Some codes require the ground around a home to be treated with insecticides to thwart termite invasions. Although this seems to work, periodic application of pesticides is not the healthiest way to address this issue. An insect-resistant coating on the foam is a safer option. Better yet, you may want to install rigid insulation product like Roxul Drain Board made from basalt rock and slag; this rock-wool board is quite insect resistant and also resists water penetration.

To reduce heat loss through the ground, more and more builders are opting for insulation applied to the interior of foundation walls. As shown in figure 2-4, insulation applied to the inside of a stem wall or the inside edge of a slab-on-grade foundation thermally isolates the foundation from the cold

FIGURE 2-4
*Heat escapes from homes through the foundation. (a) External insulation often proves to be less effective at blocking heat flows than (b) internal insulation.*

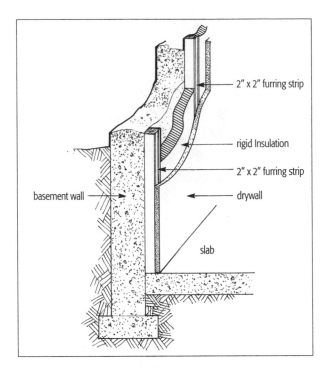

**FIGURE 2-5**
*Basement walls in new and existing construction can be insulated internally by applying rigid foam or some other insulation product between furring strips that serve as an anchor for wall board.*

**FIGURE 2-6**
*A frost-protected shallow foundation typically consists of vertical and horizontal insulation. R-values and the dimensions of insulation vary with climate. For example, the designer may opt to provide vertical wall insulation only in warmer climates, or wing insulation only at the corners in moderate climates, or vertical and wing insulation around the entire building in cold climates.*

ground surrounding a home. This strategy won't stop heat loss, but it does greatly curtail it.

When slabs are incorporated as thermal mass, as in many passive solar homes, rigid foam insulation should be inserted between the slab and the stem wall. This creates a thermal break that greatly curtails heat loss, as shown in figure 2-4. For even greater protection, many architects recommend the use of subslab insulation as well (more on this in chapter 4).

Internal insulation is also the method of choice in retrofits—specifically, when insulating a pre-existing basement (figure 2-5). Builders and designers will find a wider array of insulating products available for internal insulation strategies for at least two reasons: Insulation does not have to be waterproof, and the threat of insect damage is eliminated. However, internal insulation does not assist in damp-proofing a foundation as exterior insulation would. Therefore, water more easily seeps into an internally insulated foundation or basement wall, although a good coat of foundation sealer, perimeter drainage, and other measures, discussed shortly, reduce this problem. Furthermore, internal insulation needs to be covered with a fire-resistant material to reduce toxic outgassing in the event of a house fire.

### Frost-Protected Shallow Foundations

In recent years, a number of U.S. builders operating principally in colder regions have begun installing frost-protected shallow foundations (figure 2-6).

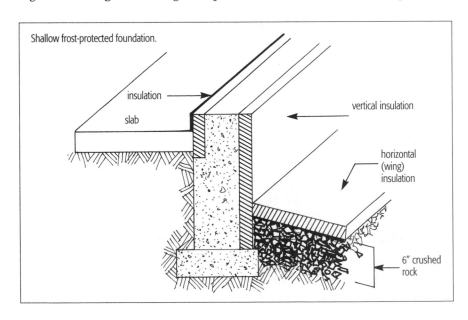

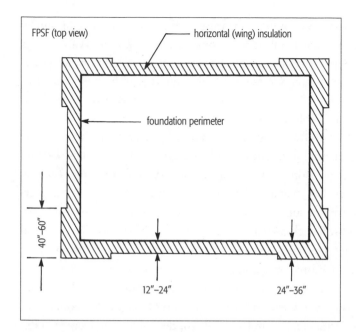

FPSF (top view)

horizontal (wing) insulation

foundation perimeter

40"–60"

12"–24"

24"–36"

FIGURE 2-7
*Top view of a frost-protected shallow foundation, showing variable wing insulation along different parts of the foundation.*

FIGURE 2-8
*Heat is captured by the frost-protected shallow foundation, reducing the temperature difference between the foundation and surrounding soil.*

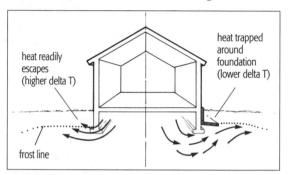

heat readily escapes (higher delta T)

heat trapped around foundation (lower delta T)

frost line

Frost-protected shallow foundations are made from concrete, although other materials can be used, and incorporate both vertical and horizontal rigid foam insulation. Also known as *wing insulation,* the horizontal portion projects laterally two to six feet from the edge of the foundation (figure 2-7). Additional insulation is typically installed at the corners of a foundation where heat loss is greatest.

Frost-protected shallow foundations create a zone of warmth around a home, trapped largely by the vertical wing insulation buried about one foot below finished grade, as shown in figure 2-8. Warmth is generated by heat from the ground and the house.

Because this technique warms the ground around a home and decreases the frost depth of the soil it permits much shallower and less expensive foundations while offering the same benefits as deeper foundations, that is, foundations dug below the frost line. The added warmth, for example, safeguards against frost heave, a condition that occurs when water in the soil around a home freezes and expands under the foundation, causing the structure and overlying walls to shift and crack. This can lead to serious structural damage.

This ring of warm subsoil also reduces the difference in temperature—or delta T—between the inside of the foundation and the surrounding earth. As delta T decreases, so does heat loss. This design therefore reduces heat loss through foundations and conserves energy. Because heat is trapped around a foundation, front-protected shallow foundations can be made much shallower than ordinary foundations—as shallow as sixteen to twenty-four inches—even in the most severe climates, places in which five to six feet deep foundations are the norm. The reduction in depth translates into significant savings in excavation, materials, energy, and labor—up to 15 to 25 percent!

## THE IMPORTANCE OF GOOD FOUNDATION DRAINAGE

Good drainage is important with any foundation, and frost-protected shallow foundations are no exception. Insulation performs better in drier soil conditions. Ensure that ground insulation is adequately protected from excessive moisture through sound drainage practices, such as sloping the grade away from the building. Insulation should always be placed above the level of the water table. A layer of gravel, sand, or similar material is recommended for improved drainage as well as to provide a smooth surface for placement of horizontal wing insulation.

National Association of Home Builders, *Design Guide for Frost-Protected Shallow Foundations*

**FIGURE 2-9**

*Notice the bed of crushed rock supporting the wing insulation in this frost-protected shallow foundation. Gravel and other measures help to drain moisture from the soil surrounding a home, which reduces heat conduction away from the building.*

Frost-protected shallow foundations were used in the 1930s in Chicago by Frank Lloyd Wright. Like many innovations in resource conservation, this technique gained popularity in Europe, particularly the Scandinavian countries. In fact, over one million homes have been built in Scandinavia with frost-protected shallow foundations since the 1950s, which is now considered standard practice in this cold climate.

Frost-protected shallow foundations are ideally suited to slab-on-grade foundations for residential structures in colder climates, but can be modified for other types of foundations—for example, a stem wall foundation, floating slab, and unvented crawl space foundation. They can also be used in commercial buildings, but are not suitable for buildings located over permafrost, permanently frozen subsoil such as that found in Alaska, Northern Canada, and Siberia. As a rule, frost-protected shallow foundations are not appropriate in areas where the mean annual temperatures fall below 32°F (0°C).

The construction of a frost-protected shallow foundation is similar to that of a conventional foundation, except for the addition of wing insulation. R-values for vertical insulation vary from R-4 to R-10, depending on average ambient temperature, while footing depth varies from twelve to sixteen inches. Horizontal insulation varies from none in the warmest climates to R-14 in the coldest. The width of the wing insulation varies as well, from twelve to sixty inches, once again depending on climate. *The Design Guide for Frost-Protected Shallow Foundations* (National Association of Home Builders) offers detailed information on vertical and horizontal insulation for different climates. You can download a copy of this publication at the Consumer Energy Efficiency and Renewable Energy Network's Web site (see the Resource Guide).

Frost-protected shallow foundations are built with water-resistant rigid foam insulation—that is, foam rated for burial. Polystyrene products are preferred. Molded expanded polystyrene (sometimes called *beadboard*) offers an environmental advantage over extruded polystyrene because it is made without ozone-depleting HCFCs and at least one manufacturer is now producing molded expanded polystyrene that is rated for underground use.

Another factor vital to the success of a frost-protected shallow foundation is proper drainage. For optimal performance, place a six-inch layer of crushed rock, sand, or gravel under the wing insulation to reduce soil moisture around the foundation (figure 2-9). The drier the soil, the less the heat loss.

The success of a frost-protected shallow foundation also relies on the elimination of cold bridges,

courtesy K-X Faswell

**FIGURE 2-10**
*Insulating concrete forms are permanent polystyrene forms filled with concrete and reinforced with steel (rebar). They use much less concrete than a conventional foundation and provide superior insulation.*

routes for the escape of heat. Be sure that materials with high thermal conductivity are isolated from the ground and air by insulation.

### Insulating Concrete Forms

To avoid the additional labor involved in insulating a standard concrete foundation and to reduce concrete use, numerous manufacturers now produce foam-insulated forms for concrete foundations. They are commonly referred to as insulated concrete forms, or ICFs. A typical example is shown in figure 2-10.

ICFs are used to build foundations and exterior walls, and come in many varieties. (At this writing there were about fifty ICF manufacturers in the United States alone.) Some manufacturers produce hollow foam blocks with interlocking edges. Assembled on a footing, the ICFs are reinforced with steel and filled with concrete. Other manufacturers produce longer, interlocking foam panels. Like the blocks, they are reinforced with rebar to increase tensile strength, then filled with concrete. Polysteel panels produced by American Polysteel Forms in Albuquerque, New Mexico, come in sizes ranging from 1 foot × 8 feet to 4 feet × 8 feet.

Unlike standard wooden or steel concrete forms, insulated concrete forms are lightweight and relatively easy to assemble. Pieces generally snap in place like toy blocks. They typically contain plastic or steel cross bridges that attach one side of the form to the other and thus resist blowout when concrete is poured into the forms.

Another feature that distinguishes ICFs from standard steel or wooden concrete forms is that ICFs are left in place after the concrete has cured. In other words, there is no need to tear down forms. The foam sandwich produces a foundation wall with an R-value of around R-30, which is very likely

**BUILDING NOTE**

Houses built entirely from Polysteel forms (walls and foundation) have reported savings of 50 to 80 percent on heating and cooling.

the highest R-value of any foundation system in use today. Because of this, they are ideal for building foundations for passively conditioned homes.

Another advantage of ICFs is that they significantly reduce the amount of concrete needed to build foundations. Reward forms, manufactured by EPS Building Systems in Broomfield, Colorado, for example, require 25 percent less concrete than an 8-inch poured wall, yet are 50 percent stronger. Moreover, ICFs are relatively easy to install and reduce the amount of lumber needed to build a home. Savings in labor and materials often render ICFs cheaper than standard concrete forms. (As a side note: whenever possible, use the expanded polystyrene ICFs, as opposed to those made from extruded polystyrene. No ozone-depleting hydrochlorofluorocarbons (HCFCs) are used in the production of expanded polystyrene.)

## Reducing Moisture Around the Foundation for Improved Efficiency

No matter what type of foundation you select, it is important to keep it as dry as possible. Water in the soil around a foundation dramatically increases heat loss. Bruce Bierup, a designer and builder of high-thermal-mass homes from Silverthorne, Colorado, underscores the point: "Wet earth around a home," he says, "acts as a giant heat sink, sucking energy away from the foundation." Moisture in wet soil and subsoil also has the potential of seeping into foundations, because concrete wicks moisture. It may then be drawn into the walls of a home, where it can dampen the insulation, dramatically reducing its effectiveness. In addition, moisture in wall cavities may cause mold and mildew, contaminating indoor air with spores. Moisture can also cause framing members and sheathing to rot, leading to structural damage.

Fortunately, numerous strategies allow a builder to reduce soil moisture around a foundation. Building on dry soil is the first and most important measure, although it is not always possible. Elevated locations are more suitable than natural depressions or swampy areas. Try to select sloped, well-drained sites. Grading—sloping the ground away from the foundation—is useful in most instances, too.

If existing drainage on a site is directed toward a home, it may be necessary to create earthen dams and alternative channels to divert water around the structure. Efforts may need to commence considerably uphill. Because surface runoff, water flowing along the surface of the ground, is enhanced when land has been stripped of vegetation, such as trees and grasses, it is helpful to replant. Vegetation acts as a sponge, soaking up moisture and reducing surface flows.

Two inexpensive but highly useful devices that reduce the accumulation of moisture around the foundation of a house are the downspout extender and flexible plastic tubing. Both are attached to downspouts and should transport rainwater and snow melt at least twenty feet from a house.

**KEEPING A FOUNDATION DRY**

• Select a dry site to build a home
• Divert surface water around a home
• Grade dirt away from the house
• Install downspout extenders, flexible tubing or underground pipe to carry water away from the house
• Install perimeter drainage

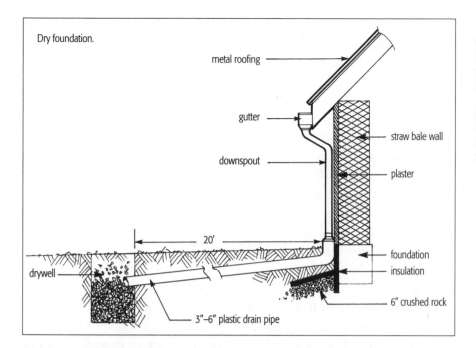

Dry foundation.

metal roofing

gutter

straw bale wall

downspout

plaster

20'

drywell

foundation

insulation

6" crushed rock

3"–6" plastic drain pipe

Catherine Wanek

**FIGURE 2-11**
*Drain pipes connected to downspouts can be used to divert water away from the foundation of a home, keeping it drier and thus reducing wintertime heat losses*

Underground drainpipes can also be installed to transport water away from a house (figure 2-11). Drainpipes are typically daylighted, that is, allowed to empty on the surface of the ground at least twenty feet away from the house. In addition, drainpipes can be routed to underground cisterns. Water that collects in the tank can be used for irrigating lawns and gardens. Rainwater and snow melt can also be transported by drainpipes to drywells, strategically placed rock-filled holes that permit water to percolate into the subsoil.

French drains running under or alongside the foundation are also useful for removing moisture from the immediate vicinity, and thus reducing heat loss. A French drain consists of a rock-filled trench with porous pipe or tile running along its bottom, which is pitched to drain by gravity away from the house (figure 2-12). Filter fabric is typically placed over the pipe to prevent sediment from clogging the openings in the pipe.

Recommended in high-precipitation regions and on poorly drained sites, perimeter drainage systems work well—so long as they have been designed and installed correctly. They are also highly recommended when walls are earth sheltered. As a side note, these ambitious forms of drainage may require professional engineering and additional excavation, but the additional costs are well worth it in the long run.

**FIGURE 2-12**
*French drains help remove moisture along foundation walls, reducing potentially damaging frost heave and costly heat loss.*

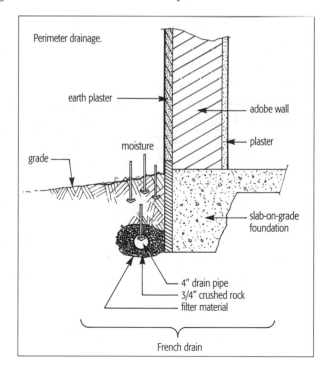

Perimeter drainage.

earth plaster

adobe wall

plaster

moisture

grade

slab-on-grade foundation

4" drain pipe
3/4" crushed rock
filter material

French drain

## INSULATION FOR WALLS, CEILINGS, AND FLOORS

A dry, well-insulated foundation greatly enhances the performance of a passive solar home, especially if the structure contains a slab that serves as a heat-absorbing/heat-releasing thermal mass. Wall, ceiling, and floor insulation also play an important part in reducing unwanted heat transfers—heat loss during the winter and heat gain during the summer—thus benefitting the homeowner year-round. As noted in the previous chapter, insulation is one of the key principles of passive design, and improvements in efficiency standards in building codes now make good insulation standard practice.

When determining the type of insulation to use for a home, the designer, builder, and homeowner must consider many factors, including (1) long-term performance, (2) source, (3) environmental impacts, (4) composition, including recycled content, (5) worker safety, and (6) the long-term health of a home's occupants. Further complicating matters, different situations require different types of insulation.

Insulation falls into four categories with specific uses: (1) loose fill, (2) blankets (rolls and batts), (3) rigid foam, and (4) liquid foam. Loose fill and blankets are typically used in attics, between roof rafters, in wall cavities in wood-frame homes, and in floors between floor joists. Rigid foam insulation is usually applied externally, for example, on roof decking, external sheathing, or foundations, although it may be used in other ways as well—for example, in wall cavities. Liquid foam is most often used in wall cavities or to seal air gaps in the building envelope, that is, the walls and roof.

### Conventional Insulation Options

Although many options are available for insulating a home, we'll begin by focusing on the most commonly used forms: fiberglass, cellulose, and rigid foam board (see table 2-1). We'll also examine mineral wool insulation, a once-popular option that is still used in some areas, and liquid foam products.

As in so many aspects of building, the choice between one insulation product and the next is rarely clear cut. Each has its advantages and disadvantages. Because many forms of insulation have undergone dramatic improvements to address legitimate concerns regarding health, safety, and environmental impacts, you may find it necessary to discard some previously held notions and biases.

*Fiberglass Insulation.* Fiberglass is the most popular insulation product on the market today and comes in two forms: blankets (batts and rolls) and loose fill. Blankets backed with paper are installed in wall cavities between studs and floor cavities between joists. Unbacked blankets are installed in ceilings, as is loose-fill fiberglass. (Note that loose-fill fiberglass insulation can also be blown into wall cavities.)

Standard fiberglass insulation products are fairly inexpensive and effective; however, they pose a couple of health risks. In fact, cancer warnings are

### FIRE SAFETY OF CELLULOSE INSULATION

The fire safety of cellulose has been challenged for years, largely by the fiberglass industry. Because wood fiber—the basic raw material in cellulose insulation—burns and glass does not, fire safety concerns are justified. However, most evidence suggests that properly treated cellulose poses no greater fire risk than fiberglass. In fact, because cellulose blocks air flow better than fiberglass, cellulose may be more effective at stopping the spread of fire.

ALEX WILSON,
"Insulation Comes of Age,"
*Energy-Efficient Building*

posted on most fiberglass insulation sold in the United States today.

One of the chief problems with fiberglass insulation is that microscopic slivers may become dislodged during handling and can be inhaled by workers. Although health concerns are disputed by the fiberglass manufacturers, studies show that these fibers may become embedded in the walls of the tiny air sacs in the lungs, known as *alveoli*. Here, the dagger-like fibers irritate the delicate lining of the air sacs, much like asbestos fibers, causing lung disease. Some studies also show that the fibers may shear DNA molecules in the nuclei of cells, which could cause mutations that could lead to lung cancer.

Although inhalation of fibers during installation can be prevented by wearing a respirator or a certified dust mask, all three principal U.S. manufacturers of fiberglass insulation have dealt with the errant fiber problem by sealing fiberglass batts in plastic film, either perforated polyethylene or polypropylene sheeting materials (figure 2-13). These products, known as *encapsulated fiberglass batts,* are popular among do-it-yourselfers and health-conscious installers. (Incidentally, the plastic also acts as a vapor barrier.)

Standard fiberglass insulation is also made with a formaldehyde-based binding agent. This binder, which holds the fibers together, outgases into the room air, affecting both installers and the occupants of a home. Besides being an irritant and causing chemical sensitivity, formaldehyde is classified as a probable human carcinogen by the International Agency for Research on Cancer.

Manufacturers have responded to legitimate concerns over formaldehyde as well. Owens-Corning, for instance, has introduced a formaldehyde-free fiberglass insulation product called Miraflex. It contains two different types of glass fiber, which expand and contract at different rates as they heat and cool. This causes the fibers to curl and twist and thus bind together naturally. As a consequence, Miraflex requires no formaldehyde binders.

Miraflex is soft and easy on the skin. It doesn't feel scratchy like ordinary fiberglass. Moreover, the fibers have greater tensile strength than those in standard fiberglass insulation. As a result, they don't break off and become airborne as readily as those found in standard fiberglass. Miraflex, therefore, not only eliminates formaldehyde binders, it protects workers from inhaling fibers.

In 2002, Johns Manville, one of the nation's leaders in insulation manufacturing, announced that it intended to replace phenol formaldehyde binder

FIGURE 2-13
*Encapsulated fiberglass batts greatly reduce health threats to workers caused by errant fiberglass fibers.*

Fiberglass (and other fiber insulations) can be mixed with binders and blown into wall cavities. A layer of mesh netting is first stapled to studs in stick-frame construction. Small holes are then cut in the netting, and the fiberglass/binder mixture is blown in. This technique ensures that air spaces around wires and pipes are well filled and thus provides a more consistent layer of insulation throughout a wall.

in all of its fiberglass insulation products by the end of the year, replacing it with an acrylic binder. This step eliminates concern over the potential health effects of formaldehyde exposure to workers, installers, and the occupants of homes.

Another advance in fiberglass insulation is the use of recycled glass in some products. This alternative costs about as much as—maybe a little less than—fiberglass batt made from virgin materials.

Manufacturers now also produce high- and medium-density fiberglass batts. In these more tightly packed batts, smaller airspaces between the fibers reduce air circulation, which increases heat resistance or R-value. Standard fiberglass blankets for 2 × 4 walls, for example, have a nominal rating of R-11 while the comparable high-density blankets are rated at R-15. Higher-density batts for 2 × 6 walls and attics are also available, and, like the 2 × 4 batts, compete well with cellulose with respect to R-value.

Higher-density batts cost about 20 percent more than their predecessors, but they provide more R-value per inch. The higher initial investment will more than pay for itself. Because of the added cost of these new products, some builders have taken to cramming oversized standard fiberglass batts into wall and ceiling spaces. For instance, a standard R-19 batt for a 2 × 6 wall can be forced into a 2 × 4 wall. The result is an R-14 wall—a wall with nearly as much insulation value as is provided by the high-density 3.5-inch batt, according to environmental building expert Alex Wilson of BuildingGreen, Inc., publishers of a variety of useful tools for builders including *Environmental Building News*. This approach can be used in closed ceiling spaces as well but requires vent spaces to prevent moisture from accumulating in the insulation. (Insulation ventilation is briefly mentioned in chapter 1 and will be discussed in more detail in chapter 3.) High-density batts do not require venting.

Fiberglass has its pros and cons. On the plus side, it doesn't shrink, won't burn, and is non-nutritive to insects. Environmentally and people-friendly products are now available, too, as just pointed out.

On the negative side, fiberglass blankets do not seal wall and ceiling cavities very tightly. Unless an installer is using encapsulated batts, a vapor barrier is required to create an airtight envelope to protect the insulation from moisture. Also, fiberglass settles and sags, resulting in a decline in R-value over time.

*Cellulose Insulation.* One of the best choices in environmentally sound loose-fill insulation is cellulose. Made from recycled newspaper and (sometimes) small amounts of cardboard shredded into a fluffy mass, cellulose is blown dry or sprayed wet (slightly damp) into wall and ceiling cavities as well as attics.

## TABLE 2-1. R-Values of Loose-Fill Insulation

| MATERIAL | R-VALUE PER INCH | USES |
|---|---|---|
| Fiberglass (low density) | 2.2 | Walls and Ceilings |
| Fiberglass (medium density) | 3.0 | Walls and Ceilings |
| Fiberglass (high density) | 2.6 | Walls and Ceilings |
| Cellulose (dry) | 3.2 | Walls and Ceilings |
| Wet-Spray Cellulose | 3.5 | Walls and Ceilings |
| Rock Wool | 3.1 | Walls and Ceilings |
| Cotton | 3.2 | Walls and Ceilings |

Loose-fill cellulose insulation is typically treated with boric acid to reduce mold, increase its fire resistance, and repel hungry insects. Some manufacturers add a moisture-activated acrylic binder that causes the cellulose particles to adhere better to reduce settling. Settling, in turn, decreases "loft," one factor that causes the R-value of various forms of insulation to decrease over time.

Not only is cellulose insulation one of the greenest alternatives, it is also fairly popular, and is giving the leader of the pack, fiberglass, a good run for its money. Fed by the massive waste of a paper-hungry society, recycled cellulose insulation manufacturing facilities are helping to put mountains of discarded newspapers and cardboard to good use. Cellulose insulation is generally a little less expensive (up to 25 percent) than fiberglass batts, too, and provides more insulation per inch (R-3.2 per inch) than standard, low-density fiberglass (R-2.2 per inch). Although it is cheaper to purchase, installation costs may be higher than for fiberglass batts.

Another advantage of recycled cellulose over fiberglass is that it is a fairly low-embodied-energy material. In addition, cellulose is a healthier product than standard fiberglass. Although blowing or spraying cellulose into wall and ceiling cavities produces a lot of dust, a certified face mask or respirator can protect workers from inhaling potentially troublesome dust.

As good as it is, cellulose insulation does have some disadvantages. Dry-blown cellulose may settle and sag, creating air spaces. It also absorbs moisture. As a result, the R-value of cellulose insulation may decline over time. Moisture can cause cellulose to mold and rot, if it remains damp for prolonged periods. Wet-blown cellulose is supposed to solve these problems, but not entirely.

*Mineral Wool Insulation.* For many years, mineral wool was the most widely used type of insulation in use in the United States, Canada, and Europe. Mineral wool insulation is similar to fiberglass, except that the fibers are manufactured from natural stone or waste from iron-ore blast furnaces, rather than glass. If natural stone, such as basalt or diabase, is used, the product is called rock wool. If iron-ore waste or slag is used, the finished product is known as slag wool. Roxul, a European company with a manufacturing plant in Canada, produces a mineral wool insulation containing equal amounts of rock wool and slag wool.

Mineral wool comes in batts and loose-fill form like fiberglass. Although mineral wool is much heavier and costs more than fiberglass, it does offers some substantial benefits. For one, it is resistant to moisture, maintaining its insulative properties when wet. It is also a better acoustic insulator and is more resistant to heat than fiberglass. Mineral wool insulation is noncombustible and withstands temperatures greater than 1,800°F (1,000°C) and actually acts as a fire barrier, slowing down a house fire and allowing more time for fire fighters to bring it under control. Fiberglass insulation, in contrast, melts at slightly over

## INSTALLING CELLULOSE INSULATION

Dry cellulose insulation can be blown into wall cavities using a plastic vapor barrier to hold the insulation in place or through holes in the drywall in walls or ceilings. In attics, dry cellulose is blown between the rafters, sometimes over them. Wet cellulose is sprayed into cavities, then rolled to trim off excess. A vapor barrier is then installed. In closed ceiling cavities, wet cellulose is carefully blown through holes in the drywall.

## RECYCLED MINERAL WOOL INSULATION

If you want to use an earth-friendly insulation batt, you may want to consider mineral wool insulation blankets made from 92 percent recycled mineral slag, ceramic tile, and rocks. Mineral wool "blankets" fit between wall studs or ceiling rafters. The manufacturer, Fibrex, Inc. in Aurora, Illinois, claims that mineral wool insulation is noncombustible and will not absorb moisture, rot, settle, or break down.

1,100°F (approximately 650°C), and cellulose burns during house fires. However, during handling small fibers can break loose from mineral wool batting, raising health concerns similar to those expressed for fiberglass.

*Rigid-Foam Insulation.* One form of insulation whose popularity has grown steadily in recent years is *rigid foam,* also called foam board or board-stock insulation. Rigid foam insulation is typically used to insulate foundations and slabs, as noted earlier in the chapter. It is also applied externally on roofs and walls and may be used as a substitute for loose fill or batts in walls, roofs, and floor cavities, although it must be tightly fitted to prevent air infiltration.

Rigid foam insulation offers significant benefits over cellulose and fiberglass. For one, rigid foam insulation products have a higher R-value per inch than blankets or loose fill materials (table 2-2). Insulation values range from about R-4 to R-6.5 (without foil facing), nearly twice the R-value of standard fiberglass and cellulose (R-2.2 and R-3.2 per inch, respectively). Some rigid foam products are ideally suited for foundation insulation because they are water resistant and can be buried in the ground on the external surface of a foundation.

Rigid foam insulation comes in three varieties: (1) expanded polystyrene (EPS), (2) extruded polystyrene (XPS), and (3) polyisocyanurate. Although most rigid insulation is made from various polymers (foam plastics), builders can also purchase additional rigid insulation made from basalt rock and slag (the same materials used to make some mineral wool insulation products). It is used to insulate foundations externally. The R-values of rigid board insulations are in table 2-2. Be sure to study your options carefully, for product content, environmental impact, and moisture resistance.

*Expanded polystyrene* (EPS), the most environmentally benign option because it is not manufactured using ozone-depleting chemicals, is a closed-cell insulation material most readers are familiar with, as it has long been used to manufacture coffee cups and packing beads for shipping. Expanded polystyrene can be molded into large sheets for insulation with R-values ranging from about 3.8 to 4.4 per inch (depending on the density of the material).

Commonly referred to as beadboard, EPS is made from polystyrene beads mixed with liquid pentane, a hydrocarbon, known as a *blowing agent.* The function of pentane and other blowing agents is that they expand to form millions of tiny bubbles in the finished product. The pentane diffuses out of the foam, and the spaces fill with air. (According to authors of *GreenSpec,* plants can be built to recover

**TABLE 2-2. R-Values of Rigid Foam and Liquid Foam Insulation**

| Material | R-Value per Inch | Use |
|---|---|---|
| Expanded polystyrene | 3.8 to 4.4 | Foundations, walls, ceilings, and roofs |
| Extruded polystyrene | 5 | Same |
| Polyisocyanurate | 6.5 to 8 | Same |
| Roxul (mineral wool)* | 4.3 | Foundation exteriors |
| Icynene | 3.6 | Walls and ceilings |
| Air Krete | 3.9 | Walls and ceilings |

* Rigid board insulation made from mineral wool.

up to 95 percent of the pentane, reducing pollution.) Air is a poor conductor of heat, so the millions of tiny bubbles effectively block heat transfer through the foam. However, air spaces in EPS can accumulate and retain water. Because water is a good conductor of heat, some form of moisture barrier may be required when using EPS to prevent this problem, especially when using this product around foundations.

Beadboard is manufactured in different densities, depending on the application. EPS manufactured for roof insulation, for example, is typically dense enough to withstand foot traffic. Wallboard insulation is much less dense. R-values vary with the density.

Anyone who has used beadboard knows that the stuff is quite brittle and is therefore not recommended for underground use (figure 2-14). To make the product more durable, waterproof, and thus suitable for burial, EPS is now being manufactured with thin foil and plastic facings (figure 2-15). Fortunately, there are alternatives to pentane-produced expanded polystyrene foam, notably a product called Insulfoam, produced by a company of the same name in Aurora, Colorado. This forward-looking manufacturer uses steam rather than pentane to generate bubbles in foam board.

*Extruded polystyrene* (XPS) is a closed-cell insulation board. Many know it as blueboard, although it also comes in a pink variety. This product is made from polystyrene and a hydrochlorofluorocarbon (HCFC) blowing agent. Although HCFCs are the much less damaging to the ozone layer than their predecessors, the chlorofluorocarbons (CFCs), the new-generation blowing agents do break down in the atmosphere when exposed to sunlight, releasing chlorine atoms that react with ozone in the stratosphere, destroying it. The good news is that, whereas a molecule of CFC destroys 100,000 ozone molecules, a molecule of HCFC only destroys about one-fifth as many—20,000 molecules. Even so, HCFCs are slated to be phased out by the year 2030.

Extruded polystyrene foam board not only damages the ozone layer, it costs a little more than EPS foam. It does, however, have a slightly higher

**FIGURE 2-14**
*Some forms of rigid foam insulation have no place in underground applications, such as this conventional polystyrene foam, also known as beadboard. After a few years, it had begun to deteriorate, reducing its effectiveness.*

**FIGURE 2-15**
*Insulfoam produces an expanded polystyrene rigid foam insulation that is coated with plastic on one side and foil on the other. This product is resistant to water and is rated for underground use. It is made without ozone-depleting CFCs or HCFCs.*

R-value than expanded polystyrene, on average about R-5 per inch. XPS also tends to be more consistent in density throughout, has a higher compressive strength than EPS, and is much more resistant to moisture.

*Polyisocyanurate* is the least environmentally friendly foam board insulation. Also known as polyiso it is a closed-cell foam made with an HCFC blowing agent, specifically HCFC-141b. Although it does less damage to the stratospheric ozone layer than the compound it replaced, HCFC-141b is the worst of the new generation (less benign) blowing agents.

Polyiso offers superior insulation properties—in fact, the highest of the rigid foam insulation materials—with an average R-value between R-6.5 and R-8 per inch. Moreover, polyiso is manufactured with various facings, for example, plastic or aluminum, which further increases its resistance to heat.

Like other forms of insulation, polyiso suffers from thermal drift, a gradual deterioration of R-value over time. Right out of the mold, polyiso foam boards are rated about R-9 per inch, but over the next two years the R-value declines to about R-7, its stable insulation value. Foil facing also adds about R-2 to the insulation.

*Some Words of Caution on Rigid Foam Insulation.* Foam board can be used to insulate many elements of a home. No matter what type of foam you use, be sure to protect it from direct sunlight, because ultraviolet radiation damages foam insulation products.

Another important consideration when installing foam board is its potential to permit air infiltration. In most homes, rigid foam insulation products are installed over external sheathing materials. Sheets must be placed close together and joints must be taped to reduce air infiltration.

*Polyurethane: Spray Foam Insulation.* In the 1970s and 1980s, many homeowners concerned with rising fuel prices opted to retrofit their homes with insulation. One product widely used during this period was polyurethane foam. Injected into wall cavities through holes drilled in the finished walls, the foam was applied in a liquid form. It quickly expanded to fill the cavity—unless it met some obstruction!

Polyurethane foam soon began to cause problems. As many homeowners quickly learned, these products were made with a binding agent containing formaldehyde. Formaldehyde released from the walls, in turn, caused numerous health problems, some very serious. In addition to containing formaldehyde, polyurethane products relied on a blowing agent known as CFC-11. It, we soon found out, is a potent ozone-destroying chemical. In addition, many found that the foam offered incomplete protection. It failed to fill cavities evenly, resulting in cold spots.

As health and environmental concerns mounted and the price of CFC-11 rose (due to a U.S. government-levied tax on the CFC designed to discourage

BUILDING NOTE

The aluminum facing in Celotex's Tuff-R insulating sheathing, for instance, adds R-2.8 when placed toward an air space in a wall cavity.

its use), polyurethane foam fell into disuse. Today, however, the product is once again on the market. Manufacturers have eliminated formaldehyde and substituted carbon dioxide gas for the ozone-depleting blowing agents. One of the leading polyurethane products is known as Icynene, a thin liquid. It is applied by a trained applicator and is sprayed in a paint-thin layer into open wall cavities (figure 2-16). Icynene adheres readily to all surfaces it contacts and quickly expands to approximately 100 times its original volume, filling cavities. Excess is trimmed off with a handsaw.

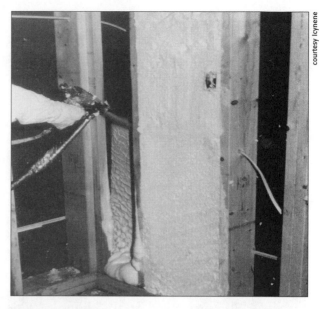

**FIGURE 2-16**
*Icynene is sprayed into open wall cavities, then expands enormously to fill the space, creating an airtight, water-resistant layer of insulation.*

Icynene provides about R-3.6 per inch. It also creates an airtight seal that reduces air filtration. Because it forms an airtight seal and is also resistant to water vapor, icynene eliminates the need for vapor barriers in standard stick-frame construction. Further adding to its benefits, icynene does not settle, sag, or shrink. Because it is structurally stable, icynene's R-value does not decline over time. Additionally, icynene production uses no CFCs or HCFCs and the foam does not outgas harmful chemicals. It offers no nutritive value to termites.

Although Icynene is typically sprayed into open cavities—stud spaces before wall board or paneling is applied—it can be applied to closed wall cavities. To do so, a carefully measured amount is poured into the cavity through a small opening in the wall. The product then expands from the bottom up, filling the voids, providing R-4 insulation per inch.

Another more environmentally friendly product is a water-blown polyurethane foam made by Resin Technology Company in Ontario, Canada. Foam Tech of North Thetford, Vermont, produces SuperGreen, a polyurethane product that contains a hydrofluorocarbon (HFC) expanding agent. Because HFCs contain no chlorine atoms, these compounds do not deplete the ozone layer.

While we're on the subject of spray-in foams, another foam product worth considering is Air Krete. Introduced in the late 1970s, Air Krete is one of the few insulation products that chemically sensitive individuals seem to tolerate—an inorganic substance containing magnesium oxide. Like other spray-in foam insulation materials, Air Krete is stable and does not shrink or settle after being applied. It supplies about R-3.9 per inch, but must be applied by trained applicators.

*Structural Insulated Panels.* While we're on the topic of foam insulation products, it is important to mention structural insulated panels or SIPs. Structural insulated panels consist of rigid foam insulation sandwiched between sheathing

**BUILDING NOTE**

Structural insulated panels (SIP) are not a new technology. They've been around since the 1950s. They're only now beginning to be widely used.

**FIGURE 2-17**
*Structural insulated panels are used to build walls, floors, and roofs.*

materials, most often oriented strand board (figure 2-17). They are used to build walls, floors, and roofs of conventional and passively heated and cooled homes. Structural insulated panels come in 2- to 12-inch thicknesses and vary in size from 4 feet × 8 feet to 8 feet × 24 feet. Even larger sheets are available for specialty applications. SIPs are assembled on the foundation and cut to accommodate windows and doors. However, most manufacturers will custom-make panels: that is, they will provide a builder with precut SIPs tailored to the house design, including door and window openings. The builder submits blueprints to the manufacturer, who produces panels and cuts them to size. Once completed, the structural insulated panels are numbered and shipped to the building site ready for assembly. Although this option costs a bit more than ordering standard panels, it minimizes cutting and trimming on site. This, in turn, reduces waste, saves time, and decreases labor costs.

SIPs are typically locked together by special joints. To make building even easier, some manufacturers produce panels with drywall mounted on the inside face and/or siding on the exterior, so a builder can assemble a complete wall with a few adroitly placed nails.

Manufacturers use three different foams in their structural insulated panels: expanded polystyrene (EPS), extruded polystyrene (XPS), and polyurethane. As noted above, the manufacture of extruded polystyrene and polyurethane uses ozone-depleting chemicals; expanded polystyrene production does not. Because EPS is also less expensive, it is generally the manufacturer's foam of choice. If the idea of using polystyrene, an oil-based product, does not appeal to you, you may want to check out Agriboard, in Electra, Texas. This company manufacturers SIPs from compressed straw and hay sandwiched between oriented strand board.

Because they are quite rigid, SIPs require little, if any, framing and, therefore, provide a viable alternative to 2 × 4 or 2 × 6 construction. SIP walls are also more airtight and energy efficient than standard stud-frame walls. R-values of the foams used in SIPs range from 4 to 7 per inch, depending on the type of foam. Therefore, a 4-inch-thick SIP (with 3.5 inches of foam) has an R value of 14 to 24. Compare that to a 2 × 4 stud wall containing standard fiberglass or rock wool insulation which produces R-values of 11 to 15, respectively. SIPs provide better insulation than high-density fiberglass and wet-blown insulation, too.

Because they lack studs, SIPs also eliminate bridging loss. Bridging losses can reduce R-values of a standard wood-framed wall by as much as 25 percent. SIPs also reduce air infiltration and accompanying heat loss, further adding to their appeal as a wall-building material in passively heated/passively cooled homes.

SIP costs vary from 10 percent less to 10 percent more than a conventional stud wall, depending on the manufacturer. Despite higher initial costs, SIPs may be cheaper than standard stud-wall construction. This type of construction requires builders to make hundreds of small cuts and assemble hundreds of pieces. That's not the case with SIPs. Because a minimum amount of cutting and nailing is involved in assembling a wall from SIPs, labor costs are reduced, provided the crew is experienced in SIP installation. Moreover, because SIPs are straight, true, and plumb, unlike 2 × 4s and 2 × 6s, builders don't waste a lot of valuable time sorting through lumber for straight pieces or discarding lots of worthless warped and bowed dimensional lumber. In addition, walls, floors, and roofs made from SIPs result in little if any bridging loss, and therefore result in a higher effective R-value than conventional wood-framed walls of similar thickness. Consequently, SIPs produce a much more energy-efficient building envelope, which greatly reduces the amount of energy required to heat and cool a home, reducing operating costs. SIPs also cut building costs by reducing the size of the back-up heating and cooling system.

SIPs are produced by numerous manufacturers and are readily available in most areas of North America. For a list of manufacturers, check out *Green-Spec* produced by the folks at *Environmental Building News*; the Austin Green Building Program's Web site; or James Hermannsson's book, *Green Building Resource Guide*. Before sending your order in, however, be sure that the structural panels you buy meet code in your area.

## Natural Insulation

Rigid foam, cellulose, and fiberglass are the three mainstays of insulation. But there are some natural materials worth considering, including cotton, wool, saw dust, straw, and straw-clay.

*Sheep's Wool.* Sheep's wool is a superb insulator. Although grazing can cause problems, for example, if sheep overgraze their range, wool insulation is a good choice for environmentally conscious builders. It has a slightly higher R-value than standard fiberglass and is produced naturally with little, if any, fossil-fuel energy.

One of wool's greatest benefits, besides being produced naturally, is that it insulates when wet, a claim that can't be made about many other forms of insulation. Wool is naturally flame resistant, too. Although wool can be damaged by moths, it contains lanolin, a naturally occurring oil that protects it

**FOAM OPTIONS**

**Expanded polystyrene (EPS)**—made from polystyrene beads and pentane; also known as beadboard.

**Extruded polystyrene (XPS)**—made from polystyrene and a blowing agent, most commonly hydrochloro-fluorocarbon (HCFC).

**Polyurethane**—made from urethane and HCFC blowing agent.

from these insects. To further safeguard wool insulation, some individuals add cedar shavings and moth balls. (Note that mothballs contain volatile organic chemicals that may be harmful to human health.)

Although the use of wool insulation may seem unusual, in New Zealand, where sheep outnumber people by a wide margin, wool insulation is common. In North America, you may be able to purchase wool insulation batts from green building suppliers listed in the Resource Guide at the end of the book. New Zealand-based Woolhouse International produces a product called Thermofleece Natural Wool Insulation, which is sold in the United States. Thermofleece contains boric acid as a flame retardant and a couple of other chemicals (polymers) to enhance its loft (its ability to spring back after compression).

*Cotton Insulation.* Another form of natural insulation worth considering is cotton. It has the same R-value per inch as cellulose insulation (table 2-1). Although not widely available, cotton insulation is manufactured in two forms: batts and loose fill. Greenwood Cotton in Rosewell, Georgia, produces a batt and loose-fill cotton insulation made from 75 percent recycled cotton-mill waste and 25 percent polyester (a synthetic). It is treated with a flame retardant but is manufactured without the formaldehyde binders found in many forms of fiberglass insulation. This product currently costs more than fiberglass batt insulation.

As environmentally appropriate as cotton may seem, you'll want to remember that cotton production is one of the most chemically intensive and environmentally harmful areas of agriculture. Soils are sprayed with herbicide to control weeds and crops are sprayed with insecticides to kill insects, often many times during each growing cycle. According to Cedar Rose Guelberth, a supplier of healthy building materials, pesticide residues in cotton insulation sometimes affect chemically sensitive individuals.

*Straw.* Straw can also be used for insulation. Some natural builders install straw bales in their ceilings to provide high levels of insulation. On the plus side, straw bales are inexpensive and readily available. However, they are also pretty heavy, with three-string bales weighing upwards of 90 pounds. As such, they're difficult to install. If you're seriously considering straw bales, don't forget that framing must be up to the challenge. Additional framing adds to the cost of a home.

Another approach tried by some adventurous natural builders is loose straw ceiling insulation. Loose straw or flakes of bales are lighter than bales and thus reduce the need to fortify roof framing. For those searching for a natural insulation, this may be a wiser and cheaper material than intact bales. In addi-

**FIGURE 2-18**
*A straw-clay wall before plastering. As a rule, natural building materials such as straw-clay, straw bale, adobe, and rammed earth do not require vapor barriers. Water vapor is allowed to diffuse out of the walls naturally and vapor barriers can trap moisture in these walls, causing damage.*

tion, in earthquake-prone regions flakes or loose straw insulation also reduces the weight above your head. (Bales could come tumbling down on top of people during an earthquake.)

Unfortunately, loose straw doesn't provide as much insulation as baled straw. In addition, loose straw poses a much greater fire hazard than intact bales. To reduce fire potential, straw can be treated with natural flame retardants such as boric acid or clay slip (described next).

Another natural insulation is a material called *straw-clay*. Straw-clay is made by mixing loose straw and a watery solution of pure clay and dirt, known as clay slip. The mixture is then packed into wall forms and left to dry (figure 2-18). The result creates a thick, insulated wall.

Some builders use straw-clay for ceilings, too. Unlike loose straw described above, straw-clay is naturally fire resistant. The clay not only protects against fire, it also retards mildew and mold growth.

## PROTECTING INSULATION FROM MOISTURE

To work well in a passively conditioned home, or any home for that matter, most forms of insulation—notably loose-fill and batt insulation—must be kept dry. Keeping insulation moisture-free requires careful design. But the first line of defense against moisture is preventive: reducing interior moisture sources. Run the kitchen exhaust fan when cooking and run the bathroom exhaust fan when showering or bathing. Cover aquariums and mulch plants. Move indoor clotheslines outside. Be sure that dryer vents open to the outside and ducts are well sealed. Be certain that heating ducts are well sealed, too.

### Vapor Barriers

Vapor barriers are also effective in preventing moisture from penetrating walls and dampening insulation. As noted in chapter 1, vapor barriers typically consist of 6-mil (millimeter) polyethylene sheeting stapled to the framing members of exterior walls and ceilings. Overlapped to further retard the escape of water, vapor barriers protect insulation, preventing its R-value from plummeting. As a result, they are an essential component of passively conditioned stick frame homes.

In addition to protecting insulation in passively conditioned homes, vapor barriers also reduce exposure to potentially harmful chemicals in building materials, such as formaldehyde in fiberglass insulation, oriented strand board, and plywood. We'll explore this subject in more depth in chapter 6.

Vapor barriers are placed on the "warm side" of a wall or ceiling of conventionally built homes (especially wood-frame buildings). In temperate and cool climates, that means vapor barriers must be located near the interior space, typically just beneath drywall or interior paneling, as shown in figure 2-19b. In hot, humid climates like the southern United States, vapor barriers

Moisture accumulating inside walls in straw bale homes could create mold and mildew and could cause the straw to decay. If dampness accumulates in an earthen wall, water can erode the material, seriously weakening the walls. In these structures, the use of breathable plasters allows moisture that moves into a wall to escape, so it doesn't accumulate in the wall. Vapor barriers are not necessary in homes constructed with natural building materials; in fact they would cause serious problems.

**BUILDING NOTE**

Drywall covered with latex paint acts as a pretty good vapor barrier, but the addition of a plastic vapor barrier adds a crucial measure of assurance, especially in an air-tight passive solar home.

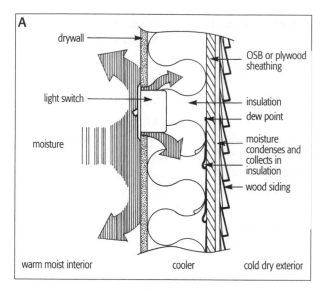

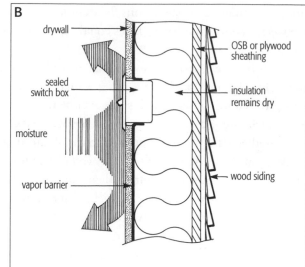

FIGURE 2-19
*Vapor barriers help reduce air movement into and through walls and thus retard moisture movement as well, protecting insulation from damage. (a) No vapor barrier; (b) vapor barrier placement for cold climate.*

## BUILDING NOTE

Movement of air into wall cavities in standard wood-frame homes can also be greatly curtailed by the airtight drywall technique, a method of installing drywall without a vapor barrier. This technique limits air infiltration by using adhesives, foam, and caulk during drywall application to seal potential leaks. Useful in production home building, the airtight drywall technique helps seal along top and bottom plates on exterior walls, around rough openings for door and windows, and in other locations as well. Special electrical boxes are used to prevent air from escaping via this route.

JEANNIE LEGGETT SIKORA,
*Profit from Building Green*

work best if placed on the outside of the wall to prevent outside moisture from penetrating into walls (figure 2-19b).

Vapor barriers may also be needed in homes with crawl spaces to prevent ground moisture from entering from beneath. Vapor barriers are also valuable under slabs, especially when slabs are used as thermal storage in passsive solar homes and when radiant floor heat is installed. In natural homes, vapor barriers aren't necessary. In fact, they can seriously damage walls (see sidebar on page 71).

Although vapor barriers are essential in closed-ceiling designs, they are generally not required in attics. The reason is that there is plenty of room for moisture entering from the rooms below to escape. If you build a house with an attic in a humid climate and need to get rid of moisture that enters the attic, you can install attic vents or a solar-powered roof vent. This device consists of a small fan that removes the moisture from an attic space year-round. Solar roof vent fans are powered by small photovoltaic (solar electric) modules.

In the winter, roof vents help to reduce moisture build-up in insulation and thus ensure optimal performance. However, they also cool the attic space in the summer, which, in turn, keeps the house cooler. This device therefore contributes to passive heating and cooling.

In closed ceilings such as a shed roof or clerestory roof, additional moisture protection can be provided by creating a small air space between the insulation and the roof decking, as shown in figure 2-20. Ridge line and soffit vents are then installed to permit air to circulate through this space and carry off moist air. Don't build without them! You may also want to consider a cold roof design (figure 2-21).

As a final note on the subject, be sure to install and use exhaust fans in bathrooms, laundry rooms, and kitchens. Exhaust dryers to the outside. And remember, latex paint over drywall helps to reduce its permeability to water vapor.

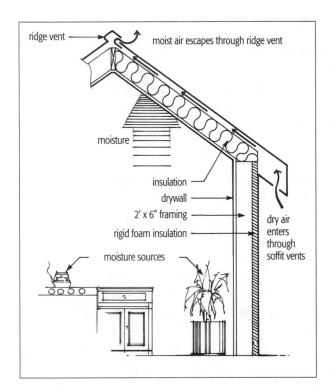

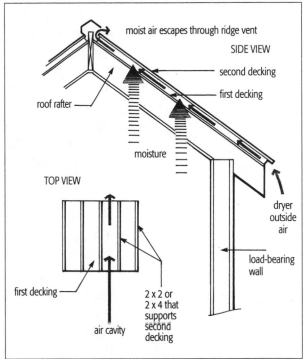

## Housewraps

In standard stud-frame houses, further protection is provided by the housewrap. A housewrap is a thin, lightweight plastic fabric made from polyethylene or polypropylene. It is stapled onto the exterior sheathing to reduce air infiltration and to prevent water striking exterior surfaces (for example, siding) from seeping into the wall. Like the fabric Goretex, however, housewraps permit water vapor to pass through. Consequently, any water that enters the wall from interior room air can escape. This protects insulation, ensuring optimal year-round performance in a passively conditioned home.

For maximum effectiveness in reducing air infiltration, seams should be overlapped and taped. Windows and doors should be flashed as well to prevent moisture from entering wall cavities.

Housewraps may or may not be required on new housing, depending on the local code and the type of external sheathing being used. They are usually required with shingles, shakes, and board sidings. According to many builders, housewraps work best in moister climates, for example, in hot, humid climates and colder moist climates, such as the Northeastern United States. In warm, dry climates, they have limited utility, as moisture is rarely a problem in such places.

## ENERGY-EFFICIENT WINDOWS: SORTING OUT THE OPTIONS

In passively conditioned homes, wise choices in north-, south-, west-, and east-facing glazing is essential to efficient year-round operation and comfort.

**FIGURE 2-20**

*Ridge and soffit vents help to reduce moisture build-up in roof insulation in closed ceiling designs, as shown here. Dry outside air enters the soffit vent in the overhang, then travels along a continuous two-inch space above the insulation. Moisture that migrates through the roof is passively drawn out of the joist cavities, exiting via the ridge vent.*

**FIGURE 2-21**

*A more expensive option than ridge and soffit vents to remove moisture from a closed roof design is the cold roof design. As shown here, 2x4s or 2x2s are nailed to the roof decking, forming channels that permit air to move between the first and second layers of decking. This vents escaping water vapor through the insulation to be vented.*

Thanks to advances in materials and design, windows have undergone a dramatic change in the last two decades from energy pores in a building's envelope to energy neutral, even net energy producers. Writing in *Solar Today*, green building expert Alex Wilson notes, "Windows have changed more rapidly than almost any other building component in the past two decades. It is now possible to let light and views into interior spaces without paying the energy penalty exacted by older glazing systems."

Unfortunately, there are so many window options that one can easily become confused. To reduce the confusion, this section explores two important topics, general window types and special energy-efficiency features of modern windows suitable for passively conditioned homes.

## Types of Windows

When considering windows, many factors come into play, including aesthetics, size, mechanics, and, of course, cost. For most people, aesthetics and cost rank numbers one and two. Although these are important, other considerations are even more important when selecting windows for a passively conditioned building. We'll begin with mechanics, that is, how windows operate, because this characteristic profoundly affects a window's energy performance.

All windows consist of three basic components: glass, sash, and frame. The glass provides for view and solar gain. The sash holds the window glass in

FIGURE 2-22
*Windows come in many forms, as illustrated here.*

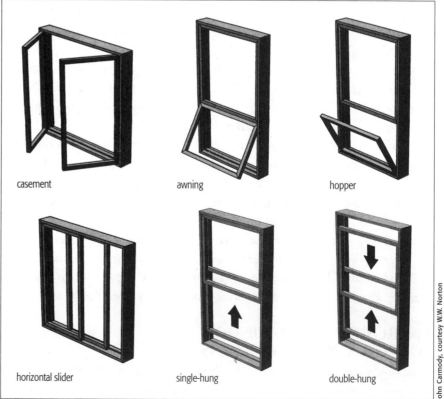

casement     awning     hopper

horizontal slider     single-hung     double-hung

John Carmody, courtesy W.W. Norton

place, and the frame holds the glass and sash in the wall. Window types are based on whether or not the sash moves, as well as the manner in which it moves.

In general, windows are grouped into two functional types: openable and nonopenable. The most basic type of window is the nonopenable variety. It is simply a piece of single- or double-pane glass mounted in a sash that is mounted in a frame. The window frame is mounted in a rough opening in a wall. All other factors being equal, fixed or nonopenable windows are the most energy-efficient option because they boast the lowest air infiltration rate of all windows. In addition, nonopenable windows are the cheapest choice.

All other windows fall into the openable category. One of the most popular is the double-hung window, consisting of two pieces of glass, each in its own sash, one above the other (figure 2-22). The sashes are offset and each slides in an independent track that is affixed to the window frame. The lower window slides up and the upper window slides down.

Single-hung windows are identical to the double-hung windows except for one feature: In a single-hung window, the upper sash is fixed. When opening the window, the lower window must be raised.

Another mechanically similar window is the slider or sliding window. They are much like double-hung windows but mounted on their sides. As such, the sash slides horizontally.

Next on the scale of complexity is the casement window. In a casement window, the sash is hinged on one side, allowing the window to swing open much like a door. Awning windows are very similar, but are hinged on the top of the window frame so the window opens outward. Next is the hopper window. It is horizontal like an awning window but hinged on the bottom and opens inward.

Clearly, your choice of openable windows is large. For the most part, there is little functional difference between various options. They all allow air to flow into and out of a house, providing natural ventilation vital for passive cooling. The casement window, however, has the added benefit of being able to act as a wind scoop, funneling breezes into a house.

Choose windows carefully and don't overdo openable windows. When tightly closed, openable windows generally permit greater air infiltration than nonopenable windows. Use only as many openable windows as you really need. Furthermore, when shopping for openable windows pay close attention to air infiltration. As discussed shortly, some are leakier than others.

## Energy-Efficient Window Design

Once you've decided on aesthetics, size, and the type of windows you'd like, it is time to consider energy performance. All three basic window components— glass, sash, and frame—influence energy efficiency of a window assembly. Although much attention is paid to glass, the sash and frame constitute about

## TEMPERATURE DIFFERENTIALS

On cold days, the temperature of the surface of interior glass of a low-e window is 10° to 15°F warmer than a double-pane window without coatings and 40 degrees warmer than a single-pane window.

25 percent of a window's total surface area, so it is important that they are energy efficient, too.

*Selecting Efficient Glass.* Fortunately for contemporary homeowners, most modern windows are designed and built for energy performance. For example, most windows are double glazed, consisting of two panes of glass separated by a small air space. It is not the two panes of glass that reduce heat loss so much, but the air space between them, because air is a poor conductor of heat.

*Inert Gases.* Nowadays, window manufacturers often fill the air space with an inert gas, typically argon. Inert gases conduct heat less well than air and add nearly one R unit to a double-pane glass window. Even better than argon is the exotic gas, krypton, although it is considerably more expensive than argon and therefore is used only in the highest-performance windows.

Unfortunately, both argon and krypton leak from windows. In a typical installation, however, studies suggest that windows lose only about 10 percent of their gas fill over a 20-year period. This slight loss translates into a very slight decline in efficiency. However, leakage may be more significant when windows are manufactured at lower elevations, for example, at sea level and installed in higher ones for instance, in mountainous areas. The pressure difference may cause seals to fail, letting the inert gases escape.

*Low-e coatings.* Another recent development of great importance in window performance is low-e glass, first mentioned in chapter 1. Low-e (or low-emissivity) glass is made by applying a thin coating—or several coatings—of clear silver or tin oxide on the glass. This transparent metallic coating allows visible light to pass through a window but blocks longer-wavelength radiation, specifically heat (infrared radiation). Low-e glass coatings can reflect up to 90 percent of the long-wave heat energy, while permitting much of the shorter-wave visible light to pass though.

As a general rule, low-e coatings increase the insulating value of a window about as much as adding one more pane of glass. Fortunately, low-e coatings are fairly inexpensive. According to window expert Paul Fisette, gas fills and low-e coatings are responsible for about 5 percent of a window's total cost. Reducing heat transfer through a window does come with another slight cost: It reduces the transmission of sunlight and, therefore, decreases solar gain.

Triple-pane windows are another low-e option. Both low-e glass and triple panes increase energy performance about 16 percent over conventional double-pane glass with no special properties.

*Soft Coat vs. Hard Coat.* Two types of low-e coatings are currently available: soft coat and hard coat. The most common, soft coat, consists of thin layers of

*Low-e glass coatings can reflect up to 90 percent of the long-wave heat energy, while permitting much of the shorter-wave visible light to pass though.*

silver oxide and anti-reflective coatings applied to the glass surface via vacuum deposition. To prevent the low-e coatings from being damaged or slowly wearing off, coatings are deposited on the inside surface of one of the panes of glass in a double-glazed window. In hot climates, low-e coatings work best if applied to the inner surface of the outer pane. This blocks heat absorption from warm outside air. In cold climates, low-e coatings are typically applied to the interior surface of the inner pane, which prevents heat inside a home from escaping.

Because soft coats can be layered, the emissivity (heat transfer) of a window can vary considerably. The greater the thickness and the number of layers, the greater the window's ability to thwart heat flow. For example, a soft coat low-e glazing on a double-pane window unit (separated by a ½-inch air space) increases the middle-of-the-glass R-value from 2.04 to 3.23, an increase of nearly 60 percent. The best possible low-e coating, called a *supercoating,* increases the R-value to 3.45.

The second major type of low-e system, hard-coat, consists of a thin layer of tin oxide incorporated into the surface of a pane of glass during manufacturing. As a result, these coatings are more durable than soft coats. Although they are more durable, hard coats do not trap heat as effectively as soft coats. The best product on the market today has an emissivity (heat transmission) of 0.2, compared to 0.15 to 0.04 for soft coats. (The lower the number the better its function.)

*Polyester Films.* Window films also reduce heat flow across window glass. Treated with a low-e coating, these thin transparent polyester films are suspended between layers of glass in double-glazed units, although they can also be applied to exposed surfaces. Thin films were developed by Southwall Technologies, an innovator in energy-efficient glass. Their product (called Heat Mirror glass) was available before low-e glass coatings.

*Window Specifications.* Low-e windows are essential for optimal energy performance of passively conditioned homes in all climates; however, requirements for north, south, east, and west glass may differ. In windows where unacceptable heat loss is anticipated, for example north- and east-facing windows, you may want to install low-e windows that retain heat inside the building. In locations where too much sun enters, for example, in west-facing windows, consider installing windows treated with heat-rejecting coatings. But what about south-facing windows?

Solar glazing requires glass that permits a lot of raw solar energy into a building, while blocking heat loss at night and during cold, cloudy periods. The designer has several options when it comes to solar glazing. Clear double-pane glass windows work well. They permit sunlight to enter unimpeded, although they also permit greater heat loss. For this reason, and because solar

### LOW-E$^2$ GLASS FOR SOLAR HOMES?

Marvin and Anderson, two major U.S. window manufacturers, use a special heat-resistant glass known as low-e$^2$ glass in all of their windows. This material provides the maximum R-value. Although this may make sense in a nonsolar home, it is not the best choice for south-facing glazing in a sun-tempered or passive solar home. Solar glazing requires glass that permits more raw solar energy to enter a building while blocking heat loss.

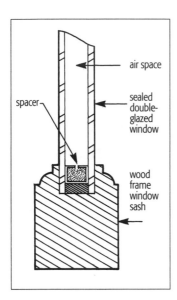

**FIGURE 2-23**
*Glazing spacers reduce conductive heat loss through window frames.*

glazing contributes such a large surface area for heat loss, many builders install low-e (heat retaining) glass along the south side of buildings as well. The high insulation value of these windows reduces heat loss in the winter.

*Warm Edges.* Many manufacturers are offering windows with "warm edges," yet another relatively new energy-saving feature. Warm edges are created by placing a nonconductive spacer between the panes of glass around the periphery of the double- and triple-glazed window units (figure 2-23). These spacers reduce heat conduction through the sash and are available for all types of windows. According to Paul Fisette, "Warm edge spacers can improve the efficiency of a window by 10 percent and can boost the edge temperature by around 5°F."

Glazing spacers also reduce condensation around window edges, which are much colder than other parts of the window. Condensation is a leading cause of call-backs on building projects. Moisture freezes and can cause ice buildup along the edges of window glass. Moisture and ice are not only annoying, they can result in serious problems. Melting ice, for example, drips onto sashes and sills, causing the finish and the wood to deteriorate. If a house generates a lot of humidity, warm edges may extend the lifetime of its windows. They'll certainly reduce maintenance.

*Energy-Efficient Window Sashes and Frames.* Another significant advance in window technology is the incorporation of improved weatherstripping materials in window assemblies. Better weather stripping results in more airtight and thus more energy-efficient windows. As Alex Wilson notes in his article on windows in *Solar Today,* "In general, hinged windows (for example, casement and awning windows) that rely on compression-type weatherstripping gaskets are tighter than sliding-sash windows." (Parentheses added.) However, manufacturers differ and so do the quality of their windows. You may find that double-hung windows from some manufacturers are tighter than casement windows from others. "So it always makes sense to examine product literature carefully," according to Wilson.

The next consideration is the type of material used to build window sashes. Metal sashes are an abomination, as they conduct enormous amounts of heat out of a house. Although the recycled content of steel and aluminum sashes and frames is a plus, metal window frames are a poor choice *unless* the window sashes are built with glazing spacers.

Wood was the first material used to build window sashes and frames. It is still widely used today. In fact, wood windows are currently the number-one selling product for residential construction in the United States. Wood windows are desirable from many standpoints. For one, wood is a renewable resource and it has a "warm feel." Wood is also a much poorer conductor of heat than metal, so wood windows tend to be fairly energy efficient. In addition, some manufacturers are now building wood frames from scrap lumber, and some are using small amounts of wood from forests certified by the Forest Stewardship Council as sustainably managed.

While wood is a superb material for building window sashes and frames, it does have some substantial drawbacks. For one, wood is not as durable as other materials and requires periodic maintenance. Rain and sun wreak havoc on exterior wood. Sunlight on south-facing windows is especially harmful to this material.

To prevent damage to wood window components exposed to the elements, manufacturers now typically install metal, vinyl, or, more rarely, ABS plastic cladding on the exterior surface of wooden window frames. This renders a window's exterior face much more resistant to weather.

Wood finishes on exposed surfaces of a window assembly inside a home—for example, sashes and frames—also take a beating from the sun and water. Sun damages finishes and water dripping from condensation on the window causes wood to swell, making it more vulnerable to further damage. Paint and other finishes peel off when wood gets wet, although factory-applied finishes can reduce this problem, as they tend to be more durable than finishes applied by homeowners or some professional painters.

Some manufacturers use vinyl, sometimes from recycled sources, instead of wood in windows. Hollow extruded vinyl sashes are comparable to wood with respect to energy performance and are therefore beneficial in passive designs. Vinyl windows are extremely durable and long lasting. Some types of vinyl are resistant to ultraviolet light and are advertised as virtually maintenance free. In addition, vinyl windows seem to be fairly price-competitive with other windows. Because vinyl is impregnated with pigment, window frames require no painting. (In fact, it can't be painted!) Many of these features make vinyl an ideal choice for passive designs.

Vinyl windows have been around since the early 1970s and are popular in Europe and Canada. They are the number-two product in the United States. Despite their popularity, vinyl windows do have some disadvantages. For one, vinyl fades in sunlight and becomes brittle with age. Because it can't be painted, fading may decrease the aesthetics of a home over time. (Soft Scrub or other similar products will revitalize the color of vinyl, reportedly bringing it close to its original hue.)

Vinyl windows are made from polyvinyl chloride (PVC). Vinyl chloride, the raw material from which PVC is made, is a known carcinogen. Vinyl chloride poses a danger to workers, although most factories have substantially reduced exposure to their employees. I've never seen any evidence to suggest that it is harmful to homeowners.

Another serious drawback of vinyl is that it expands and contracts significantly in response to changes in temperature (so it is said to have a very high coefficient of thermal expansion). Over time, notes Wilson, "the expansion

*Vinyl expands and contracts more than wood, aluminum, and glass. This can cause the window frame to develop cracks that leak air and water.*

**Metal**–excellent conductor of heat and thus a poor choice of materials unless windows are fitted with glazing spacers.

**Wood**–poor conductor of heat and thus a good choice from an energy standpoint. They do tend to be damaged by weather. Be sure windows have exterior cladding.

**Vinyl**–fairly resistant to sunlight and highly resistant to water. If designed properly, they offer long, reliable service.

**ABS plastic**–resistant to sun and moisture and offer long, reliable service if designed properly.

**Fiberglass**–costly but resistant to weather and reliable.

and contraction . . . can loosen seals, cause cracks at corners and on flanges, and lead to premature failure." This can cause leakage, which decreases the efficiency of windows. Because of these potential problems, vinyl windows are frequently designed and manufactured in ways that prevent damage, and failures are reportedly rare. "If you choose vinyl frames," says Paul Fisette, "specify heat-welded corners. They hold up best over time."

Mikron Industries of Kent, Washington, has taken vinyl one step further by combining virgin and recycled vinyl with recycled wood fiber, creating a composite that is now being used by other companies to build sashes for windows and patio doors. Unlike traditional vinyl window sashes, wood/vinyl composite is paintable.

Another similar product, but relatively new, is ABS. ABS plastic (or acrylonitrile butadiene styrene) is extruded with another resin to produce highly weather-resistant windows. A wide range of dyes can be mixed with the plastic, many much darker than those available for vinyl windows. Although ABS plastic has an expansion coefficient similar to vinyl, good design can minimize potential problems.

One additional product worth considering is the fiberglass-frame window. Fiberglass offers many of the benefits of vinyl. For example, it has low embodied energy and is made primarily from silica, one of the world's most abundant resources. Fiberglass windows are extremely strong, too. Moreover, fiberglass has a much lower coefficient of thermal expansion, much closer to glass. Thus, fiberglass windows hold up better than other synthetics and some people think that fiberglass may become the window frame of the future.

Despite these advantages, fiberglass is made using a fairly toxic resin. It probably doesn't present a hazard to the homeowner, because it evaporates by the time the window is shipped. However, the resin does pose a threat to workers.

Fiberglass windows are not as widely available as wood or vinyl windows—and they're more expensive. They must also be painted. At this writing, I am aware of only one manufacturer, Owens Corning, that is making fiberglass windows with insulated sashes.

*Assessing Energy Performance of Windows.* When buying windows, it's easy to become overwhelmed by the details of different models, especially when it comes to the energy performance required for an efficient passively conditioned home. Fortunately, most manufacturers provide information on the energy efficiency of their window assemblies.

For years, window manufacturers were left to their own devices when it came to testing and reporting on the efficiency of windows they manufactured. In 1989, however, window manufacturers joined forces to launch a nonprofit organization, the National Fenestration Rating Council or NFRC. The NFRC establishes standards for testing windows to ensure consistency; it also establishes standards for reporting the results of tests for the convenience of consumers. The NFRC also publishes a booklet that lists this data on various windows, *NFRC Certified Products Directory*.

Four measurements are typically reported on windows: (1) U-value, (2) air infiltration, (3) the solar heat gain coefficient, and (4) visual transmittance. Only U-value, defined below, is required.

*U-Value: Heat Loss.* U-value is a measure of heat transmission through a material. It is the opposite of R-value, which you will recall from chapter 1, is a measure of the resistance of a material to heat flow.

U-value is calculated by dividing R-value into 1 (U-value = 1/R). Thus, an R-value of 2 would yield a U-value of 0.5. A window with an R-value of 3 would have a U-value of 0.33. The lower the U-value, the lower its heat transmission.

U-values for windows are often reported for the middle of the glass as well as the entire assembly, which takes into account heat loss around the edges. As a rule, U-values of smaller than 0.3 (which is an R-value of 3.3 or better) are required for passively conditioned homes. As a side note, Heat Mirror glass (described earlier) can achieve U-values of 0.17 or lower. Remember that U-value can be increased by installing insulated window shades, curtains, or rigid insulated shutters, as discussed in chapters 1 and 3.

*Air Infiltration.* The second consideration in window efficiency is air infiltration. Primarily resulting from leakage between the sash and the frame, air infiltration is determined by the type of window and the quality of construction. Generally, the lower the quality, the greater the air infiltration. As a rule, a window that opens permits more air infiltration than a nonopenable window. Among openable windows, however, some designs permit more air infiltration than others. Double-hung windows and sliding windows generally permit slightly more air infiltration than casement windows and awning windows because hinged windows fit more tightly to the frame than double-hung windows and sliders. But the difference is slight and well-made double-hungs are often acceptable. The key to knowing the quality is checking the posted air leakage ratings.

Air leakage is measured in cubic feet of air per minute per square foot of window surface (cfm/ft²). Look for windows with certified air leakage rates of less than 0.30 cfm/ft², recommends window expert Paul Fisette.

*Solar Heat Gain.* Solar heat gain must also be taken into account when specifying windows for a passive solar home. Window manufacturers report this as the solar heat gain coefficient or SHGC, a term that replaces a less-useful measurement with a rather confusing name, shading coefficient.

*The U-value or U-factor is used to determine heat loss through window assemblies. The lower the number, the higher the insulating value.*

*Well-designed windows have durable weather-stripping and high-quality closing devices that effectively block air leakage.*

PAUL FISETTE, "Understanding Energy-Efficient Windows," in *Energy-Efficient Building*

*Air leakage siphons about half of an average home's heating and cooling energy to the outdoors each year. Air leakage through windows is responsible for much of this loss.*

PAUL FISETTE, "Understanding Energy-Efficient Windows," in *Energy-Efficient Building*

Solar heat gain coefficient is actually a ratio of solar heat gained by a window to the total amount of solar energy striking the window. Put another way, it is the fraction of solar radiation falling on a window that is transmitted through the glass as heat. It tells you how much of the solar energy striking a window is permitted to enter a home and takes into consideration various sun angles and the shading effects of the window frame. The SHGC of windows varies between 0, meaning no solar gain, to 1, indicating 100 percent solar gain.

South-facing glazing should have a high solar heat gain coefficient with sufficient U-value to conserve heat at night and during cloudy periods, and low infiltration for maximum efficiency. The colder the climate, the more solar gain is needed, and the higher the SHGC should be. Paul Fisette recommends SHGCs for hot climates under 0.4; for intermediate climates, between 0.4 and 0.55; in cold climates, greater than 0.55.

*Visual transmittance.* Another factor to consider when buying windows is the visual transmittance (VT), the amount of light a window transmits. Visual transmittance is reported as a percentage of the light that would pass through an open hole in the wall the same size as the window. Tinted windows permit 15 percent VT, while clear glass permits up to 90 percent VT. For most of us, a window with 60 percent VT appears clear, and one below 50 percent appears dark.

## Other Performance Issues

In addition to the four measurements just described, a few other factors are relevant to the overall energy performance of windows.

*Strategic Placement.* For optimal performance of a passively conditioned home, choose openable windows that have the best U-values and the lowest air infiltration rates. You can also enhance the performance of a home by minimizing the number of openable windows and by strategically placing openable windows in ways that permit maximum air flow. This strategy not only reduces air infiltration, it eliminates unnecessary expense by minimizing the number of openable windows. As noted in chapter 1, air flow can also be promoted by employing an open floor plan design.

Although reducing openable windows is possible in open designs, it may not be wise to reduce openable windows in more traditional house designs—that is, homes with numerous isolated rooms. In these instances, hallways, walls, and doors can substantially obstruct air flow and reduce natural ventilation. If you're designing a home with a more conventional "closed design," you will probably need to install a couple of openable windows in each room. Choose the most airtight models you can afford.

*Blocking Ultraviolet Radiation.* Windows that block ultraviolet radiation are also vital in some solar designs to protect carpets, drapes, and furniture, although visible light also fades fabric.

Most windows on the market today reduce UV penetration by 75 percent or more. Some manufacturers list a window's ability to reduce UV damage using the Krochmann Damage Function. Lower numbers are better.

*Size of Windows and Installation.* As a rule, windows generally lose more heat around the edges. Thus, the larger the ratio of glazing area to perimeter, the less heat is lost. The smaller the ratio, the greater the heat loss. Divided windows, for instance, consist of numerous double-pane sections, each with a huge amount of edge in relation to total glazing. The result is excess heat loss, compared to a single, double-pane insulated glass unit of the same size. If you want the divided light look, your best bet is to purchase windows with an applied grill.

Proper installation is also vital to reducing infiltration. Be sure to fill the gaps between the window frame and rough framing with insulation, caulk, or foam before installing window trim.

## What is the Most Efficient Window You Can Buy?

Most homes are built with high-performance windows. The economic payback period for such windows is usually between two and ten years, based on energy savings. The slightly higher initial investment in high-performance windows, however, may also be partly offset by installing smaller, less expensive back-up heating and cooling systems.

Whereas the economic payback from good-quality windows is both immediate and long term, comfort gains occur immediately and last a lifetime. Moreover, high-quality windows may last longer and require less maintenance, saving time, money, and hassle.

For approximately $1000 more (for a typical home), you can purchase super-high-performance windows. They consist of multiple glazings (two or more layers of glass), krypton gas filler, and multiple low-e coatings or multiple suspended films. The center-of-the-glass R-value is about R-12, but the overall window including all sources of heat loss performs at about R-7—approximately three R units better than the best high-performance windows.

## View vs. Solar Gain: Achieving a Good Working Balance

Although window quality is important, so is window placement. Where windows are placed, however, is rarely based solely on energy losses and gains. In fact, it is safe to say that most people value the other amenities that windows offer—view, safety, and lighting—more highly than potential solar gain. The goal in designing a passive solar home that will perform well year-round is to

**TABLE 2-3.**
**Embodied Energy of Common Building Materials and Products**

| MATERIAL | EMBODIED ENERGY IN MJ/KG (million joules per kilogram) |
|---|---|
| Bales straw | 0.24 |
| Adobe block (traditional—mud and straw) | 0.47 |
| Concrete block | 0.94 |
| Concrete | 1.0 to 1.6 |
| Concrete (precast) | 2.0 |
| Hardwood timber, kiln dried, rough sawn | 2.0 |
| Softwood timber, kiln dried, finished | 2.5 |
| Cellulose insulation | 3.3 |
| Plaster board | 6.1 |
| Cement | 7–8 |
| Plywood | 10.4 |
| Fiberglass insulation | 30.3 |
| Steel (virgin) | 32 |
| Carpet (nylon) | 148 |

From: Andrew Alcorn, *Embodied Energy Coefficients of Building Materials.* Wellington, New Zealand: Centre for Building Performance Research, 1998

achieve a good working balance between aesthetic considerations and energy performance. It is the responsibility of architects and builders to explain such matters to their clients, which isn't always easy.

Individuals interested in taking an active role in designing, even building, their own homes can also consult with a knowledgeable solar engineer or solar designer to determine window placement and optimum glazing ratios. However, some fairly user-friendly computer programs exist, such as BuilderGuide software and ENERGY-10 software, discussed in chapter 7, that allow one to experiment with different window configurations—that is, different types of windows, different window placements—and then to measure the consequences of these choices. Another useful tool is RESFEN, an inexpensive computer software program developed by the Lawrence Berkeley Laboratory. It is used to assess window function in homes. This program helps a designer of passively conditioned buildings minimize energy loss, maximize comfort, control glare, and maximize daylighting. It also assists in selecting windows for specific applications, for example, north-facing and west-facing walls. The software is available from the NFRC, which is listed in the Resource Guide. Proper window ratios are discussed in chapter 3.

## SUPER-LOW-ENERGY BUILDINGS: A COMPLETE ENERGY PACKAGE

While energy-efficient design is vital to successful passive conditioning, there are other aspects of energy that designers and builders should bear in mind. One of them is the embodied energy of the materials that go into building a home. As noted in chapter 1, embodied energy is the energy required to produce and distribute a product from mine or forest to the store shelf. To create a home that requires as little energy as possible, be sure to carefully consider the embodied energy of the materials that go into it. To obtain information on the embodied energy of building materials, log on to www.arch.vuw.ac. NZ/cbpr/index_embodied_energy.html. Here you will find a report by Andrew Alcorn from the Centre for Building Performance Research at Victoria University of Wellington, New Zealand. Table 2-3 lists the embodied energy of some of the most common building materials.

Low-energy homes also require the installation of low-energy appliances, electronics, and lighting. Front-loading washing machines, for example, use approximately one-third of the energy of a conventional top loader and half the water—cutting down on energy on both accounts. Compact fluorescent light bulbs use approximately one-fourth of the energy of standard incandescent lightings. As noted in the introduction to the chapter, they also affect cooling and heating loads.

Combined with conventional efficiency measures such as reduced air infiltration, superinsulation, and high-performance windows, low-embodied-energy building materials and low-energy appliances, electronics, and efficient lighting provide opportunities to create a super-low-energy building.

When buying appliances or heating and cooling equipment for a passively conditioned home, look for those that carry the Energy Star label. Energy Star is a program launched by the U. S. Environmental Protection Agency and Department of Energy. It began with computers and other electronic equipment, then spread to appliances, such as refrigerators and washing machines. The Energy Star label ensures that the product, whether it is a new computer or refrigerator, is energy efficient. In 1996, the EPA took the ideas that seemed to work so well in appliances and electronics into home building.

In order to obtain an Energy Star seal of approval, new homes use at least 30 percent less energy than would be required by a home built under the Model Energy Code, once thought to be a fairly advanced residential energy code and one that few states and municipalities have adopted. The Energy Star certification program is a voluntary effort between builders and the EPA. That is, builders are free to choose the materials and techniques needed to achieve energy savings. Energy performance is the goal.

To qualify for Energy Star rating, a home has to be certified by an independent third party, a home energy auditor who runs a series of tests on the building and checks building details, such as insulation and types of windows, to see if the home's energy performance meets the rigorous energy criteria.

When buying an Energy Star home, you can be assured you have purchased a building that will be efficient to heat and cool and comfortable to live in. When building your own home, you may want to hire an architect and a contractor who participate in the Energy Star program—or, better yet, exceed Energy Star efficiency recommendations.

Energy-efficient construction delivers more value for lower costs, but it can also qualify you for special lower-interest loans. PHH Mortgage Services and other companies, for instance, offer slightly lower interest rates for Energy Star homes—for example, about one eighth of a point lower than mortgage for standard homes. Therefore, even if an Energy Star home costs more, the homeowner comes out ahead.

As new Energy Star builders and lenders like to remind customers, first cost (initial cost) is truly not the bottom line. We must take into account operating and maintenance costs when determining the true full cost of a house. Paying a little more can reap huge economic benefits.

Energy is not the only issue when designing for sustainability, however. The use of green building materials, graywater recycling, roof catchment systems for domestic water, native vegetation, and a host of other measures all factor into the equation. I'll summarize them in the last chapter. These subjects are discussed in more detail in my book *The Natural House* (also published by Chelsea Green).

*Insulation, high-performance windows, reduced air infiltration, the use of low-embodied-energy materials, and energy-efficient appliances, lighting, and electronic equipment are all components of a low-energy building program.*

# PASSIVE SOLAR HEATING: REGION-SPECIFIC DESIGN

**MY SECOND HOME** was built by a contractor who had grand ideas about helping reduce energy use in homes. It was his first passive solar home. I was its second owner.

The house, which is shown in figure 3-1, was lovely to look at and breathtaking inside. It had 16-foot vaulted ceilings. I fell in love with it the instant I walked in the front door, as did my former spouse. I was especially excited about this house because it employed three different passive solar design features: direct gain, an attached sunspace, and a thermal storage wall, all of which are discussed in this chapter. The builder had oriented it properly, insulated the envelope well, and provided adequate mass—or so we thought. Even though the house was in the trees, it had a decent solar "window" with full

FIGURE 3-1
*The author's first true passive solar house. Although this home was one of the best the author looked at during his search, it turned out to have many problems, most notably overglazing and undermassing.*

We all love high ceilings, but they can be a design nightmare in colder climates. Heat rises away from occupants where it is needed most, and is difficult to move back down. Much of this heat escapes through the ceiling, resulting in cooler night-time temperatures than you may prefer. High ceilings also mean there's more volume to heat during cold, cloudless periods. That increases heat bills and pollution.

access to the sun from 9 or 10 AM to 3 or 5 PM, depending on the time of year (and therefore the angle of the sun).

As well thought out as this home was, though, we soon discovered that it had some significant flaws. Too many skylights and some large, west-facing sliding glass doors that afforded a view of the 14,000-foot Mount Evans resulted in excessive solar gain during the summer. The builder also unknowingly exceeded recommendations for south-facing glass, causing the house to overheat in the winter as well. I often walked around in shorts and a T-shirt during the dead of winter, and still felt as if I were about to spontaneously combust. Moreover, the air inside the house was unbearably hot and dry.

Overheating was a problem in this house in the summer, fall, and winter! But there were other problems, too. The vaulted ceilings permitted heat to rise and dissipate through the skylights and the roof. So at night, this lovely little solar house that baked us in the day, chilled us down like fish in a freezer.

The builder had anticipated the problem of rising hot air and had installed two fans in plenums in the walls to gather the heat and transport it to the lower level of the house. This was a great idea, but the fans were undersized and thus ineffective. I replaced them with larger models, which were still ineffective. They barely made a dent in the problem. A ceiling fan we installed didn't help much, either.

Making matters worse, the builder had installed inexpensive sliding glass doors that leaked excessively during the winter, so at night they produced a bone-chilling draft.

The builder made another costly error. He had failed to install a vapor barrier and a roof venting system. A couple years after we moved in, during the house's fifth year of existence, the shingles on the roof began to buckle. When the roofer arrived to fix the problem, we found that the decking was soaked from water vapor escaping through the ceiling. The insulation was damp as well, which reduced its effectiveness.

As time went on, we discovered additional problems. One was that the direct-gain space was virtually unusable during daylight hours. The light was too intense to watch television, read, or work at a computer.

Over the years, I came to realize that my first inclination about this passive solar home being a great place to study solar design proved true. Unfortunately, it turned out to be a case study in well-intentioned design disasters. Fixing the mistakes consumed a lot of time and cost a great deal of money—about $15,000—for example, to install Plexiglass over skylights and sliding glass doors, to re-roof the house, and to install ridge and soffit vents to prevent moisture from building up in the insulation.

## PASSIVE SOLAR HEATING

Passive solar heating systems provide space heat for a wide range of buildings from homes to banks to schools to dentist offices. Unlike other solar energy systems—for example, those that produce domestic hot water for homes—a passive solar heating system contains only one moving part: the sun. (Actually, the sun's movement across the sky is due to the Earth's rotation.) It travels in a daily arc across the sky that changes by season. In the summer, the sun carves a steep path across the sky. In the winter, its arc is low in the sky (figure 3-2). In the intervening seasons it travels an intermediate path.

Passive solar home designers take advantage of the sun's variable course, using south-facing glass to permit sunshine to enter and properly designed over-hangs to regulate solar gain, permitting the sun to enter when it is needed and blocking its intrusion during the rest of the year. Passive solar designers use other standard elements of a building—walls and floors, for example—to store solar heat and release it over time, providing warmth.

Sunlight contains energy of many different forms, such as visible light, heat, ultraviolet radiation, X-rays, cosmic rays, and gamma rays. For our purposes, visible light and heat (infrared radiation) are the most important. In a passive solar home, visible light entering south-facing glass (solar glazing) is absorbed by thermal mass within the structure and converted to heat. The fate of this heat depends on a simple laws of physics: heat always flows from warm to cold. Thus, as the surface of thermal mass warms up, heat tends to migrate inward toward the cooler interior. Heat also radiates into the air around it, if it is cooler, or is stripped away by cooler air flowing over its surface. All three forces, conduction, radiation, and convection, help distribute heat within a building.

During daylight hours, visible light produces heat that charges the mass wall and warms a home's interior. At night or during cloudy periods, when room temperature falls below mass temperature, heat flows back into the room from thermal mass. In well-designed and well-constructed passive solar homes, very little human intervention is required beyond opening and closing window shades. Nor are any mechanical devices required to ensure comfortable indoor temperatures, and it is for this reason that the process is referred to as "passive."

In most climates where heating is required, passive solar design can easily provide 25 to 80 percent of a home's annual heat requirement, known as its *heating load*. To do so, *a passive solar house must be designed and built correctly*. Proper design begins by following the fourteen principles outlined in chapter 1. Among other things, these principles emphasize the importance of

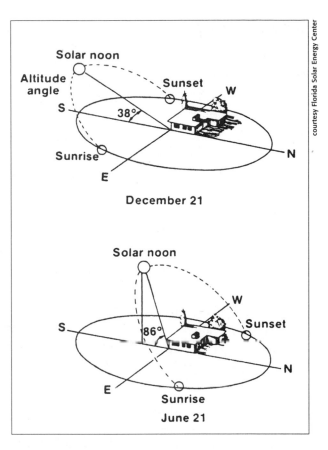

FIGURE 3-2

*The sun cuts a low arc through the sky in the winter in Florida (shown here) and elsewhere and is thus able to penetrate south-facing windows to provide heat. During the summer, the sun cuts a high arc and thus very little sunlight can penetrate south-facing windows.*

**FIGURE 3-3**
*The sun-tempered solar home gains about 25 to 30 percent of its heat from the sun. This goal is achieved by orienting the east-west axis of the home to true south and shifting some of the windows to the south side.*

choosing building sites wisely, orienting homes toward the south, and designing buildings to ensure the sun's unfettered entrance during the heating season. They point out the need to build homes that are airtight and energy efficient. Installing the proper amount of thermal mass and locating mass properly for optimal comfort and maximum solar gain are also vital to successful passive solar design. Of course, integrated design is also crucial. In this chapter, we will explore the specifics of passive solar design aimed at creating homes that function optimally in different regions, providing year-round comfort with little, if any, backup heat. We will begin by surveying two major design options—sun-tempered homes and true passive solar homes. Next, we will explore a system for regional design of passive solar homes that takes into account the heating requirements of various regions and solar availability—the amount of sunlight available for passive solar. In subsequent sections, we focus on the specifics of three passive solar designs: direct gain, thermal storage systems, and attached sunspaces. I'll present guidelines for designing each type, as well as modifications required to make each one work in mild, moderate, and severe climates.

## SOLAR DESIGN OPTIONS

Passive solar buildings fall into two broad categories: sun-tempered and passive solar.

### Sun-Tempered Designs

Sun-tempered buildings are the simplest and least expensive passively heated buildings. Sun-tempered homes are achieved by orienting the east-west axis of a house to true south and by modest increases in the number of windows on the south side of homes, which usually entails a shift of glazing from other sides of the house—and therefore is no additional cost (figure 3-3). In a sun-tempered design, south-facing glass should not exceed 7 percent of the total floor area. In other words, in a 2,000-square-foot home, the south-facing window glass should not exceed 140 square feet.

Sun-tempered designs are the least costly passive solar option. They do not require any additional thermal mass. Incidental mass in floors, walls, ceilings, and furniture is generally sufficient to prevent overheating.

Sun-tempered designs do not require insulation beyond the requirements of local building codes, provided that local building codes are adequate. Following or exceeding the guidelines of EPA's Energy Star program or the International Energy Conservation Code boosts performance with little extra cost.

Sun-tempered homes typically cost the same as conventional homes, but reap fairly substantial benefits in comfort and reduced energy costs over their lifetime. With very little extra effort, these homes can satisfy up to 20 to 30 percent of the annual heating load from passive solar, depending on local conditions, house design, and construction details (including insulation). As such, sun-tempered homes are ideal for builders and developers who want to reduce the energy demand of their houses without making the slightly higher financial commitment to true passive solar design.

## Passive Solar Designs

True passive solar designs provide substantially more heat than sun-tempered designs. To do so, designers place a greater percentage of a home's windows on the south side of the building and reduce the number of windows on the north, east, and west sides. To prevent overheating and to provide a means of storing heat to ensure greater long-range thermal comfort, passive solar designs require much greater use of thermal mass than sun-tempered designs (figure 3-4). These changes boost the amount of heat generated by the home, which can easily climb to 50 to 80 percent or more of the annual demand. (The range depends on climate, the availability of sunshine, design, and construction.)

In this book, I will focus primarily on passive solar home designs, examining three options: direct gain, attached sunspaces (isolated gain), and thermal

Sun-tempering is the simplest and least expensive passive solar strategy. It is achieved by orienting a house to true south and modest increases in windows on the south side of a home. No additional thermal mass is required, only the free or incidental mass in standard building components, for example, framing, wall board, and furnishings.

PASSIVE SOLAR DESIGN STRATEGIES

Passive solar designs concentrate glazing on the south side of the house combined with thermal mass to store heat and prevent overheating. South-facing glass ranges from 7% to 12% of the total floor space. Three types of passive solar designs are encountered: direct gain, isolated gain, and indirect gain.

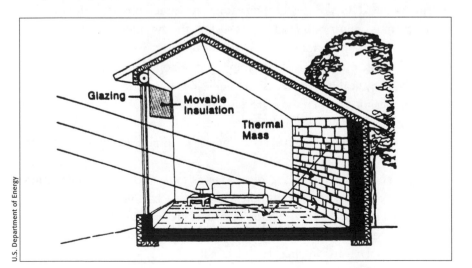

U.S. Department of Energy

FIGURE 3-4
*Direct-gain passive solar homes rely on south-facing windows to permit sunlight to enter during the heating season. Sunlight is converted to heat inside the living space, heating it directly, hence the name direct gain.*

## HEATING AND COOLING DEGREE DAYS

Heating degree days is a measure that tells us how many days each year, on average, the outside temperature falls below 65°F. At this temperature, internal heat sources (lightbulbs, people, and appliances, for instance) and solar gain in non-solar homes will result in internal temperatures around 70°F, deemed to be a comfortable interior temperature. Cooling degree days is a measure of how many days each year, on average, outside temperature exceeds 65°F.

## HEATING AND COOLING DEGREE DAY SOURCES

To obtain heating degree day data for your location, log on to www.nws.mbay.net/hdd.html. To obtain cooling degree day data for your location, log on to www.nws.mbay.net/cdd.html. (Data is based on thirty-year averages for years 1971 to 1990.)

storage walls (indirect gain). Before examining the three types of passive solar design and how to modify them for optimal performance in different regions (reflecting differences in solar availability and heating loads), we need to examine two vital aspects of region-specific design: heating requirement and solar availability.

## REGION-SPECIFIC DESIGN: HEATING REQUIREMENTS AND SOLAR AVAILABILITY

Designing a passive solar home for a particular region requires a knowledge of outdoor temperature during the heating season to estimate heating requirements. Rather than using average temperatures, however, most designers rely on heating degree days as their guidepost. (So does the software I'll introduce you to in chapter 7.)

Heating degree days is a measure of how many days each year, on average, the outside temperature falls below 65°F. At this temperature, internal temperature in a house usually hovers around 70°F, as explained in the sidebar.

To understand how the number is derived, consider a simplified example. If the outside temperature is 64°F for one full day, it is considered a 1-degree day. If the outdoor temperature is 60°F for a day, it is considered to be a 5-degree day. If the outside temperature averages 0°F for two weeks, the result is 910 heating degree days (65 degrees × 14 days = 910 heating degree days). The higher the degree heating days, the colder the climate.

Figure 3-5a is a map of the U.S. displaying heating degree day zones. Not surprisingly, heating degree days increase from south to north. As an example, northern Minnesota is rated at 10,000 heating degree days. Southern Florida is rated at 500 degree days.

As noted above, the heating degree day measurement indicates general heating requirements—how much solar heat or conventional heat a home will require to produce comfortable indoor temperatures. Obviously, the higher the heating degree days, the greater the heating challenge. In passive solar design, architects must adjust various elements such as the amount of insulation, thermal mass, and south-facing glazing to ensure comfort in different regions.

While we're on the subject, it is important to note that maps of the United States showing heating degree days should not be used for design purposes. Regional variations don't show up on nationwide maps and may result in rather dramatic differences in heat requirements. To design a passive solar home, you will need to obtain specific data on heating degree days in the area in which you will be working.

With knowledge of heating requirements, one next needs to determine available solar energy, to assess its potential to satisfy heating requirements. Figure 3-5b is a map of the United States that illustrates average daily solar energy. Once again, don't use U.S. maps to design a home. Obtain local data or state maps that display variation in solar availability.

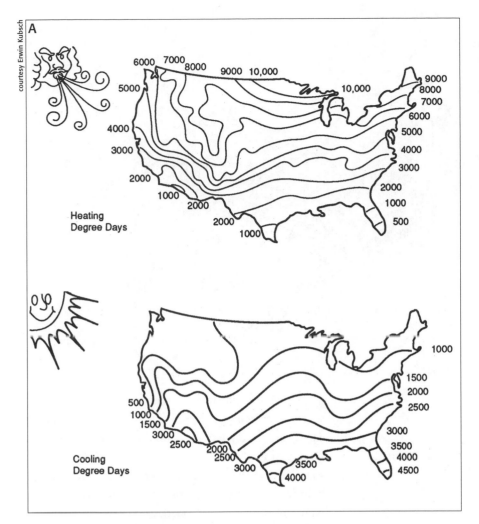

A

Heating
Degree Days

Cooling
Degree Days

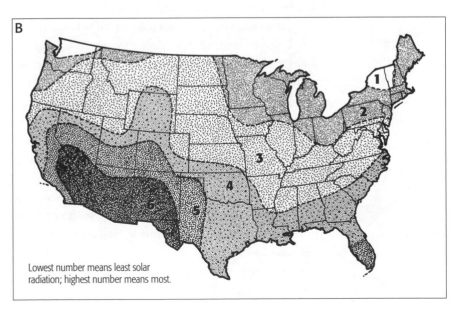

B

Lowest number means least solar
radiation; highest number means most.

**FIGURE 3-5**
*(a) Heating and cooling degree days indicate the heating and cooling challenges of an area. Be sure to use data from state maps for designing your home. (b) Average daily solar radiation. This information helps a designer determine solar availability. Again, be sure to use data from state maps for design.*

FIGURE 3-6
*Overlaying maps of the daily solar radiation and degree heating days helps the designer plot a successful strategy for passive solar heating.*

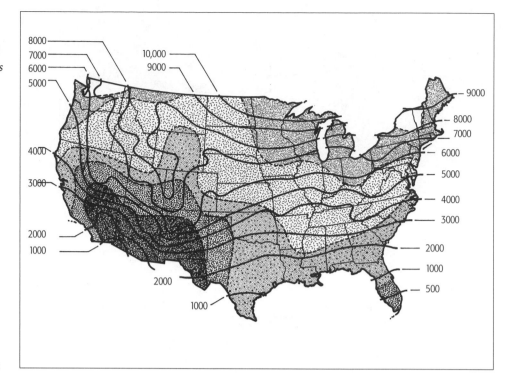

FIGURE 3-7
*Solar radiation in California. This map shows that there is considerable variation in average daily solar radiation within a state.*

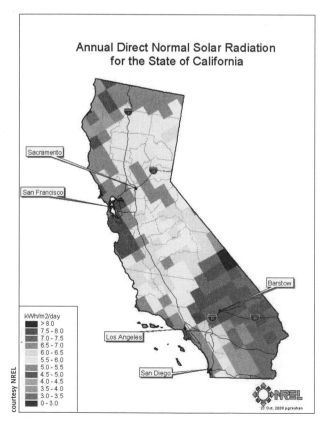

Annual Direct Normal Solar Radiation
for the State of California

Comparing heating requirements with available solar energy helps a designer determine how much heat can be provided passively from the sun (figure 3-6). As illustrated in chapter 7, both degree heating days and solar radiation are taken into account when designing a home and running performance calculations either the painfully slow, old-fashioned way—by hand—or the faster, more convenient way—by computer.

It should be clear from the map of solar radiation and degree heating days that a home in a cold but sunny climate such as eastern Montana obtains more solar heat than an identical home in a cold but cloudier climate, for instance, northern Maine, New Hampshire, and Vermont. However, a knowledge of heating requirements and available solar resources allows a designer to formulate effective compensating strategies to boost solar reliance in such instances — for example, increasing insulation or solar glazing. By taking such measures, a home in a less sunny environment can be made to perform as well as a home in a sunnier locale.

Fortunately, some relatively simple worksheets and some sophisticated software programs are available to help you custom-design buildings for specific regions. They are discussed in chapter 7. However, be sure to use data for solar radiation from your area (figure 3-7).

With this understanding in mind, we next turn our attention to the three main passive solar designs, beginning with direct-gain systems.

## DIRECT-GAIN PASSIVE SOLAR HEATING

Direct gain is the most widely used passive solar heating strategy. Appropriate for mild to moderate climates, direct gain is a relatively simple and straightforward approach. It involves the use of south-facing windows to permit low-angled sun to enter a home (figure 3-8). Sunlight absorbed by interior surfaces is then converted to heat to warm interior spaces day and night. How much solar glazing is sufficient?

### Solar Glazing in Direct-Gain Systems

In direct-gain passive solar homes, solar glazing should range between 7 and 12 percent of the total floor space. Depending on the climate and solar availability, this level of glazing could provide 50 to 80 percent of the annual heat requirements, perhaps even more.

Solar glazing options in direct-gain systems are many. In most homes, vertical windows in south-facing walls provide an avenue for sunlight to enter a home. Skylights also permit solar gain when located on south-facing roofs. Let's examine these options in more detail.

*South-Facing Windows: Vertical vs. Tilted.* South-facing windows may be tilted (angled or sloped) or vertical (figure 3-9). In most passive solar designs, vertical glass works sufficiently to gather sunlight for heating, and is the product of choice among designers. However, tilted glass offers some advantages

*In direct-gain passive solar systems, window space should not exceed 7 to 12 percent of the total floor space. In this "more-is-better world," be careful to resist the temptation to overglaze!*

FIGURE 3-8
*A direct-gain passive solar home. Sunlight streaming through south-facing windows is converted to heat inside the home, directly heating the interior space. This beautiful passive solar home in Morrison, Colorado, was designed by Colorado architect James Plagmann.*

over its counterpart. For one, tilted glazing permits greater solar gain during the winter months when the sun is lowest in the sky. During this period, sunlight strikes more or less perpendicular to the glass, minimizing reflection and maximizing solar transmission. The result is greater solar gain—that is, more sunlight enters the house and more heat is generated from the sun.

Because it is not shaded by overhangs, titled glass also permits more sunlight to enter throughout the rest of the year, and may be a better option in colder climates. Year-round solar transmission is also more conducive to growing food and other plants in planters located along the south wall.

Although tilted glass has its benefits, the problems it creates can be quite significant. While it increases wintertime solar gain and lets more light into a home for better plant growth, tilted glass may result in heat gain in the off-season (spring, summer, and fall). This can lead to serious overheating, turning a solar home into a furnace—not what you need on the Fourth of July! Moreover, one solar engineer noted that "any time a window assembly is off vertical, it will eventually leak, especially if it is exposed to extreme temperature swings." Expansion and contraction of the glass and frames caused by fluctuating temperatures compromises window seals. Water and air leaks occur as a result. Water leaks are annoying and can lead to damage. Air leaks decrease the performance and comfort of a house. In addition to these problems, angled glass is more difficult to shade than vertical glass and local building code may require tempered glass in such applications, which is more costly.

Opinions on tilted glass are split, although not evenly. The majority of passive solar designers recommend against it. I agree that ordinary vertical glazing is almost always a better year-round choice. In the winter, vertical glazing often offers energy performance on par with tilted glass as the low-angled sun reflects off snow cover, which increases solar gain, according to the Sustainable Buildings Industry Council (SBIC).

*Clerestory Windows.* While vertical and horizontal wall glass permit sunlight to penetrate deeply into a house, some designers use clerestory windows to deliver sunlight even deeper into a building, for example, onto north walls, as

**FIGURE 3-9**
*South-facing glass can be vertical or tilted. Most designers prefer vertical glass, for reasons explained in the text.*

both photos courtesy Solar Survival Architecture

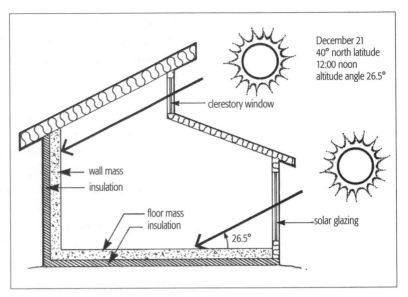

December 21
40° north latitude
12:00 noon
altitude angle 26.5°

clerestory window

wall mass

insulation

floor mass

insulation

solar glazing

26.5°

shown in figure 3-10. Clerestory windows increase solar gain and facilitate daylighting. Although they are typically vertical windows, tilted glass clerestories are also occasionally used. In some homes, clerestories are installed in a sawtooth pattern, as shown in figure 3-11, which can increase solar gain and daylighting. (In such instances, careful design is required to ensure adequate roof drainage.)

Clerestory windows should be placed in front of intended mass walls, usually at a distance approximately 1 to 1.5 times the height of the wall to ensure maximum contact with the thermal mass. For example, if the north wall is a 10-foot-high mass wall (insulated on the outside), the clerestory windows should be 10 to 15 feet in front (south) of it.

**FIGURE 3-10**
*Clerestory windows permit sunlight to penetrate the interior of a home, striking mass deep within the structure. They also assist in daylighting, however, if not covered at night by insulated shades, they can lose a significant amount of heat.*

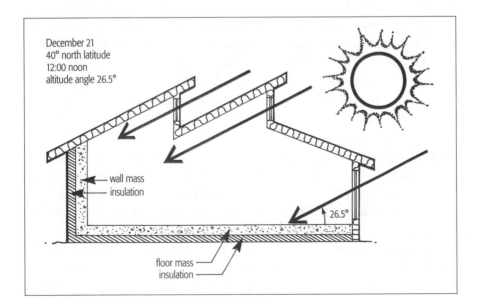

December 21
40° north latitude
12:00 noon
altitude angle 26.5°

wall mass

insulation

26.5°

floor mass

insulation

**FIGURE 3-11**
*Clerestory windows in a sawtooth pattern on commercial buildings help to deliver sunlight deep into the structure.*

Clerestory windows offer other benefits in addition to deeper sunlight penetration. For example, they may reduce glare and ensure greater privacy, both problems with south-facing wall glass. They are also useful in homes in which first-story south-facing glass is obstructed by trees, nearby homes, or other structures.

Although clerestory windows offer important benefits, they should be used judiciously because they can become serious energy liabilities: They can provide an avenue for substantial heat loss. Although clerestories can be covered with insulated shades, that's not a very practical option because they are usually twelve to fifteen feet off the floor and shades are therefore difficult to access. As a result, clerestory windows often remain uncovered at night or during cloudy periods, thus losing significant amounts of heat.

Clerestories also admit a great deal of light at night. Lights from street lights, neighbors' homes and even moonlight can make it difficult for lighter sleepers to get a good night's rest.

Clerestory windows are also associated with elevated (vaulted) ceilings, which by necessity increase the volume of air that must be heated. I've had considerable experience with clerestory windows and vaulted ceilings in the two passive solar homes I've lived in since 1987, and personally would recommend against these features in cold climates. If your design calls for clerestories, be sure to use the most energy-efficient windows. You may also want to install motor-driven window shades so the clerestories can be insulated at night.

*Go for Quality.* No matter what type of solar glazing you use, go for quality. Look for high-performance windows, well-made windows with low U-values and low rates of air infiltration, as discussed in chapter 2. Double glazing and gas fills help reduce heat loss and are almost always your best bet, even in warmer climates. (They help to keep a house cooler in the summer months as well.) Triple-paned windows may also be advisable in extremely cold climates.

Remember, too, that south-facing glazing takes a tremendous beating. If you opt for wooden windows, be sure that the surfaces exposed to the sun are protected by metal cladding with anodized paint. Chapter 2 offers a detailed discussion of window options for optimal year-round performance with minimal maintenance.

*Skylights.* Skylights, mentioned earlier, represent yet another choice for solar heat gain but only when placed on south-facing roofs of passively heated homes (figure 3-12). Bear in mind, however, the steeper the south-facing roof, the greater the wintertime heat gain. The flatter the roof, the less solar gain during the dead of winter when the sun is lowest in the sky.

Skylights are very popular in all types of homes, not just passive solar structures. Unfortunately, skylights present many of the same problems as titled glass. For example, they let in too much sunlight during the summer, which can cause overheating. Skylights are also rather costly to shade. Although they can be shaded to prevent overheating in the summer and insulated to protect against heat loss during the winter, accessing shades can be difficult, and shades and window insulation are very expensive. The more difficult they are to operate, the less likely a homeowner will attend to their opening and closing.

If improperly designed or installed, skylights can leak, causing considerable damage to roof materials, insulation, and ceilings, although newer flashing and installation techniques have helped to cut down on this persistent and costly problem. Moreover, skylights provide an avenue for heat to escape from interior spaces in the winter. A square foot of skylight glass loses about 24,000 Btus per year.

As Peter van Dresser points out in his book *Passive Solar House Basics,* "the nighttime heat loss from skylights usually exceeds the heat gain during sunny days unless arrangements are made for insulating closures." To reduce this problem in my house, my son Forrest and I invented insulated slider panels on a track system for skylights in our growing area (see figure 3-13); we're thinking of applying for a patent on this product. Made from rigid foam insulation with an R-value of 7 and covered with cloth fabric, the insulated panels fit into a simple but effective wooden railing system mounted on the ceiling. They're slid over the skylights at night or during cloudy weather to cut down on heat loss. Without them, the skylights lose approximately 1.4 million Btus per year.

When selecting skylights, be sure to purchase high-quality models with high-performance glazing. Skylights are available in much cheaper, lighter materials than glass—for instance, plastic. Although they are less expensive, plastic skylights do not perform thermally as well as glass skylights.

A far better option is the solar tube skylight and other similar models (figure 3-14). Solar tubes consist of flexible tubes that extend from the ceiling through the roof. Equipped with a light-gathering lens mounted on the roof and an internal reflective surface, these devices transmit large amounts of diffuse light from a small surface area. The result is enormous light transmission with minimal heat loss. Solar-tubes, The Sun Tunnel, Tubelites, and other

Catherine Wanek

**FIGURE 3-12**
*Skylights on south-facing roofs increase solar gain and provide daylighting. They can however, result in significant heat loss at night or during cloudy days.*

**FIGURE 3-13**
*Insulated solar panels invented by the author and his son Forrest (at age 9).*

courtesy Sun Tunnel

**FIGURE 3-14**
*The solar tube skylight permits sunlight to enter a room without the huge heat loss from a conventional skylight.*

products take up a fraction of the space of conventional skylights. Inside, they look like ceiling lights.

## Nonsolar Glazing

Effective design of a direct-gain passive solar home also requires attention to nonsolar glazing—that is, windows on the other sides of a home. Two aspects are particularly important: glass allotments and types of windows (openable vs. nonopenable). For more on windows see chapter 2.

In a direct-gain passive solar house in most climates, north- and east-facing glass should be minimized, each accounting for no more than 4 percent of the floor space. West-facing glass should not exceed 2 percent of the total floor space. These numbers are often altered to accommodate view or for safety—for example, to afford views of the yard to detect intruders and for egress (escape) during fires. However, changes in glazing allotment almost always result in a penalty, most often overheating in the summer.

When it comes to window placement, however, a good working knowledge of local climate is crucial. In cooler climates, for example, summer overheating is less of a problem than in warmer climates. Therefore, installing west-facing glass a little above the recommended level may result in very little penalty. And remember, too, that shading, special glazings, and other measures can compensate for deviations from the general guidelines on window placement.

In warmer climates with high cooling loads, a mix of north- and south-facing glass generally results in lower energy use in the summer. In such instances, reduced south-facing glazing cuts down on solar gain while additional north-facing windows release heat. The net result, if properly done, is a cool interior.

*Glazing for Ventilation.* Windows play an important part in passive cooling, a topic we'll explore in more detail in chapter 5. For now it is important to remember that when designing a passively conditioned home be sure that some of the solar glazing is openable for natural ventilation. Openable windows on the north side of the house, and possibly the east and west sides as well, are useful in creating cross-ventilation.

In two-story solar homes, a combination of openable windows located on the first and second stories helps to ventilate a home, by taking advantage of the natural stack effect (or chimney effect) caused when warm air rises to the ceiling. In the summer, heat escaping through open upper windows draws cool air into the house from windows on the north or east sides of the building, creating a continuous cool breeze. Remember that openable windows leak more and cost more than fixed (nonopenable) windows, so don't go overboard on the openable ones. A few strategically placed openable windows can create adequate ventilation.

## Insulation and Thermal Mass

A well-insulated home is a prerequisite for the success of a direct-gain system. Well-insulated means well-insulated walls, floors, ceilings, foundations, and windows to retain solar heat. Please refer to chapter 2 for detailed discussion of these topics. (Later, I'll explain how insulation should be adjusted for region-specific design.)

Thermal mass is also of great importance in direct-gain systems. As noted earlier, increasing the amount of glazing over the 7 percent limit requires additional thermal mass—that is, thermal mass beyond incidental mass—to accommodate the increase in solar gain. Thermal mass prevents overheating and stores heat, which is released passively when indoor temperatures fall below mass surface temperature. Mass therefore helps to provide more heat and greater thermal stability. Exceed this glazing limit or skimp on mass and you'll likely create serious problems; one of the most troublesome will be wintertime overheating.

But mass must also be well-positioned for high performance in a direct-gain passive solar house. As a rule, mass should be fairly evenly distributed throughout the space heated directly by the sun to provide uniform comfort. Even distribution works better than concentrating mass in one area. That said, the more mass you can locate in the direct path of incoming solar radiation the better. You should also seek ways to cast light onto mass located in the back of a home. Clerestory windows, mentioned earlier, shine light on interior or north-side mass walls.

"The principle at work here," says Ron Judkoff, director of the Buildings and Thermal Systems Center at the National Renewable Energy Laboratory, "is that surface area is more important than thickness for relatively low-conductivity materials such as masonry. Masonry is only active for the first couple of inches. For high-conductivity materials such as water, it is fine to concentrate mass."

As noted in chapter 1, for many years solar advocates assumed that thermal mass inside a house should be dark-colored. Dark-colored mass in direct contact with the incoming sunlight absorbs a lot more visible light than a lighter-colored mass and produces more heat. Experience, however, has revealed a small problem with this logic: While some of the heat produced by darker-colored thermal mass is absorbed by the material, a significant amount

### DON'T FORGET SUMMERTIME COOLING

Window shading is vital to the success of most direct-gain passive solar homes. Ron Judkoff, director of the Buildings and Thermal Systems Center at the National Renewable Energy Laboratory, notes, "Most climates up to about 7,000 degree-days really need shading in the summer to avoid excessive cooling loads"—that is, to prevent overheating. Window shades are particularly useful in reducing solar gain in the late summer and early fall. The warmer the climate, the more important shading can be. External shades generally work better than internal shades, as noted in chapter 1.

### POSITIONING YOUR MASS

As a rule, mass should be fairly evenly distributed throughout the space heated directly by the sun in a direct-gain system. This works better than concentrating mass in one area. Nonetheless, as a rule, the more mass you can locate in the direct path of incoming solar radiation the better.

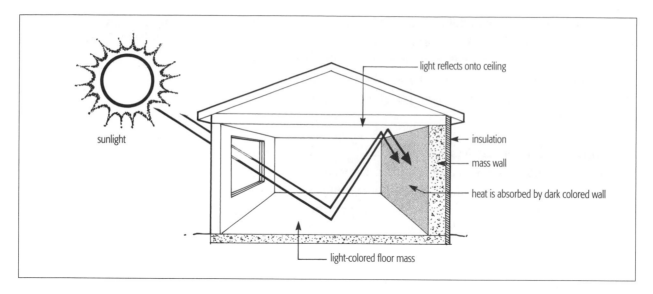

Labels in figure:
sunlight
light reflects onto ceiling
insulation
mass wall
heat is absorbed by dark colored wall
light-colored floor mass

FIGURE 3-15

*Sunlight is often reflected off light-colored mass toward the south side of a passive solar home onto darker, heat-absorbing mass situated deeper in the structure.*

can be transferred to the surrounding air. The result is a hot zone, a region of the house that suffers from uncomfortably hot air.

Rather than choose dark colors for all direct-contact mass, many architects now prefer a reflection and absorption strategy (figure 3-15). That is, they use lighter colors up front and darker colors in the back. The lighter-colored mass near windows converts some sunlight to heat, to be sure, but it also reflects a good portion of the sunlight into the interior of the building. Here, it strikes deeper, darker-colored mass, which absorbs the light and converts it to heat. By reflecting light inward to darker-colored mass surfaces in living spaces, a designer can effectively surround occupants with warm, heat-radiating walls.

While mass and insulation are generally provided by distinctly different components of a home, some building materials and wall systems provide both simultaneously. Log homes, for instance provide both mass and insulation, as explained in the box on the facing page.

## Glass-to-Mass Ratio

One of the most important design details of a direct-gain solar home is the glass-to-mass ratio—that is, how much mass is incorporated in a building in relation to the solar glazing. Follow the guidelines in this section carefully to prevent overheating.

When sizing mass in relation to solar glazing, remember that the first 7 percent of the glazing in a passive solar home is accommodated by incidental thermal mass, such as gypsum wall board, framing, floors, and furniture. (You will recall that 7 percent refers to a proportion of square feet of glass to total square feet of floor space, and is the upper limit of a sun-tempered design.)

When glazing exceeds the 7 percent value, additional mass is required. Three glass-to-mass ratios are useful (figure 3-16) in determining how much.

*Deep internal mass will pick up heat from warmed room air, but it absorbs less heat than if it is in direct contact with sunlight.*

Log homes offer many benefits to homeowners. Log homes possess a rustic, back-to-the-earth look, for example, that appeals to many people who have grown tired of the rectilinear world of conventional frame construction. But how do they rate as potential passive solar homes?

You've probably heard conflicting stories about the energy efficiency of log homes. Some people say that wood is "not a particularly good insulator" and that log home walls compare poorly, even against standard frame homes with fiberglass batt insulation in the walls. Other sources proclaim that wood is "nature's best insulator" and that log homes actually perform as well as or even better than conventional wall systems.

As with many issues, the truth is somewhere in the middle. The claim that wood is one of nature's best insulators is true. Wood has an R-value four times greater than concrete, six times greater than brick, and fifteen times greater than stone. Comparing wood to concrete blocks, bricks, and stone,

however, is rather meaningless. Most homes in the industrial nations are built from 2 × 6 or 2 × 4 studs with fiberglass or cellulose insulation. When log walls are compared to stud walls, using R-value as the sole means of distinction, log walls fail miserably. A six-inch white pine log, for example, has an R-value of about 8 and a western red cedar log of the same diameter is a measly R-6.5. Compare that to a 2 × 6 stud wall with fiberglass insulation, which is about R-19, or with cellulose, which is about R-22, and you can see the concern many people have about log structures.

But don't close the book on the subject just yet. Wood also has thermal mass and stores heat. When the thermal mass is taken into account, researchers have found that log walls perform as well or slightly better than conventional insulated frame walls with respect to heat loss. When the U.S. National Bureau of Standards performed tests on log walls, they found that a six-inch log wall equaled or exceeded the energy performance of any other type of

wall during all seasons tested except the dead of winter. In the winter, insulated framed walls won by a small margin. The thermal mass made up for the difference in R-value.

Jim Cooper, author of *Log Homes Made Easy* and numerous articles on log homes, however, notes that "Too many people see the tag 'energy efficient' as a license to ignore sensible energy conservation measures. They run amuck with energy-inefficient cathedral ceilings, glass in the wrong quantity and the wrong place, and disregard for the role of house siting in maintaining energy efficiency." Don't fall into that trap with this or any building technology.

So, if your heart is set on a passive solar log home, the structure will work. Be sure that ceiling, floor, and foundation insulation are sufficient, and add additional mass if you exceed the 7 percent glazing limit.

Adapted from Daniel D. Chiras, *The Natural House: A Complete Guide to Healthy, Energy-Efficient, Environmental Homes.*

The first ratio relates solar glazing to floor mass in direct contact with sunlight. In such cases, each square foot of solar glazing over the 7 percent mark requires 5.5 square feet of *uncovered* and sunlit (directly illuminated) floor mass.

The second ratio relates solar glazing to floor mass not in direct contact with incoming solar radiation, but in the same room. In this case, 40 square feet of uncovered and "unlit" mass accommodate one square foot of solar glazing. (The efficiency of floor mass falls off dramatically when it is covered by carpeting.)

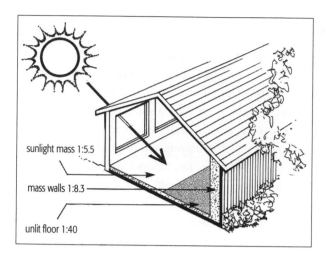

sunlight mass 1:5.5

mass walls 1:8.3

unlit floor 1:40

**FIGURE 3-16**
*Glass-to-mass ratios for floors and walls.*

The third ratio relates solar glazing to wall mass. In a room being warmed directly by sunlight, you'll need 8.3 square feet of wall mass for each square foot of solar glazing over the 7 percent limit. According to the Sustainable Buildings Industry Council, it doesn't matter whether the wall mass is directly in contact with the sun or not. However, not all experts agree with this assessment, noting that wall mass in direct contact with sunlight absorbs more heat.

Both floor mass and wall mass should be four to six inches thick. Any thicker, notes the SBIC, and the mass has little effect in absorbing more heat. Once again, some designers disagree and call for much thicker mass to provide greater thermal stability and longer periods of solar heating between recharges, as noted in chapter 1.

How do you determine how much wall and floor mass you need? Let's consider a simple example to illustrate the process.

Suppose you are building a 2,000-square-foot home with 240 square feet of solar glazing (12 percent of the floor space). Because of the 7 percent rule, 140 square feet of the solar glazing will be "carried" by incidental mass (7% × 2,000 = 140). The remaining 100 square feet of solar glazing will require additional mass, in the walls or the floors or both.

To calculate the amount of uncovered floor mass in direct contact with sunlight, multiply 100 square feet of solar glazing by 5.5 square feet of floor mass per square foot of solar glazing. The answer is 550 square feet of uncovered sunlit floor mass. In other words, to prevent overheating and to minimize indoor temperature variation, the house will need a 20-foot by 27.5-foot section of uncovered floor mass—for example, tile on a concrete slab—that is illuminated directly by incoming solar radiation. It must be four to six inches thick.

If your plans accommodate half of the additional solar glazing, you will need to make up the difference either by additional floor mass in the direct-gain space, but not in contact with the sun, or by wall mass, or some combination of the two. In this example, the remaining 50 square feet of solar glazing will require 2,000 square feet of uncovered floor not directly illuminated by the sun (50 square feet of solar glazing × 40 square feet of floor mass per square foot = 2,000 square feet) or 415 square feet of wall mass (50 square feet of solar glazing by 8.3 square feet of wall mass = 415 square feet of wall mass). Obviously, some combination of the two would work.

## REGION-SPECIFIC DESIGN OF DIRECT-GAIN SYSTEMS

With these basics in mind, we can turn our attention to designing a direct-gain passive solar home for a particular region. Although direct-gain passive solar systems are appropriate for mild to moderate climates, the design remains the same throughout its range. Four factors change over this range, however, to fine tune the system: (1) the amount of south-facing glass, (2) the amount of thermal mass, (3) the level of insulation, and (4) the size of the overhang.

## Solar Glazing

One of the trickiest factors in creating a region-specific design is solar glazing. In passive solar, the greater the need for heat, the greater the solar glazing requirements.

Simple as it sounds, this generalization can get a designer or builder into trouble. During the winter, excessive glazing can result in overheating, while at night or during cloudy periods, too much glazing can lead to excessive heat loss and uncomfortably cold temperatures. Overglazing can also lead to over-heating in the summer, especially if the house is not properly aligned to the sun.

As a general rule, solar glazing should fall within the 7 to 12 percent range of the glass-to-mass ratio discussed on page 102. If you are building in a warmer climate, you may only need 8 or 10 percent glazing. If you are building for a colder climate, you may need 12 percent glazing. Solar glazing allotments can be boosted beyond 12 percent, according to the Sustainable Buildings Industry Council, by combining solar design features—for example, adding a thermal storage wall to a direct-gain passive solar home. In such instances, the total solar glazing of all design features should not exceed 20 percent of the floor space. For nonsolar glazing, follow guidelines presented earlier (see page 100).

## Thermal Mass

The fourth adjustment for region-specific design is thermal mass. A passive solar design is only as effective as its mass. As noted earlier, mass should be fairly evenly distributed throughout the direct-gain space. The more mass placed in direct contact with sunlight or diffuse light, the better. Remember to color mass surfaces carefully to distribute light into the interior of a direct-gain space. As a rule, mass should be lighter up front and darker in back.

Note, too, that interior finishes over mass can greatly reduce heat absorption. Carpeting, for instance, virtually negates heat absorption by floor mass. Plaster or stucco has little effect on heat gain in wall mass, as it allows heat to conduct from the surface to the interior. In contrast, drywall over wall mass will reduce heat absorption unless it is tightly affixed to the wall. Recommendations for the thickness of masonry mass and the amount of mass in relation to glazing are found on page 100.

Mass is sized in relation to solar glazing. So once you've determined your solar glazing, you can calculate how much mass you need. Use the floor plan to determine where mass could go. Then run the calculations to determine how much mass of each type you'll need to accommodate the solar glazing.

**SOME NOTES ON THERMAL MASS**

Thermal mass converts sunlight to heat energy, which it absorbs. It also absorbs heat from air, but this mode of energy transfer is less efficient than that occurring when mass is struck by sunlight. According to Steven Winter Associates, "storage mass that is heated only indirectly by warm air from the living space requires roughly four times as much area as the same mass in direct sun to provide the same thermal effect." As a general rule, wall mass generally performs better than floor and ceiling mass.

## Insulation

Envelope insulation is important in all direct-gain systems. The colder the climate, the greater the level of insulation. In cold climates, such as those of Maine or Minnesota, R-30 to R-40 wall and R-50 to R-60 ceiling insulation are common in high-performance passive solar homes. Design homes to reduce or eliminate thermal bridging loss through framing members in walls, ceilings, roofs, and floors. As you may recall from chapter 2, thermal bridging is the loss of heat through framing members and foundations. It occurs through conduction. (Consult chapter 2 for some ideas on ways to reduce thermal bridging.)

When it comes to insulating a house designed to make use of solar heat, most architects and designers I've talked to exceed insulation requirements stipulated by their local building codes. Although local codes have been upgraded over the years, many designers and builders opt for EPA Energy Star recommended insulation levels or better. Others insulate at or above the International Energy Conservation Code now adopted by many jurisdictions. Chapter 7 examines computer software that will allow you to experiment with designs with different levels of insulation. You can run calculations longhand as well, but that is a time-consuming and laborious process.

Window insulation is especially important in direct-gain systems. Like wall and ceiling insulation, the colder the climate the more important the window insulation becomes. Combined with high-performance windows, insulation can help reduce heat loss during evening hours and cloudy periods and ensure that windows provide a net energy gain.

Window insulation takes several different forms—for example, internal or external shutters and shades. Well-made insulated shutters provide a higher level of insulation than shades, although they require more effort on the part of the homeowner, especially when they're installed on the outside of a house.

Floor insulation is also crucial in direct-gain designs, especially if the floor serves as thermal mass. When designing for all but the warmest climates, it is best to isolate the thermal mass in floors—for instance, in slabs—from the underlying ground with rigid foam insulation, as explained in chapter 2. Cold earth surrounding a passive solar home draws heat out of the building, reducing its efficiency and comfort. High moisture content in the soil will have the same effect. Be sure to design a foundation and slab so that they remain dry through peripheral drainage, grading, and other means. Again, chapter 2 describes ways to keep foundations dry and discusses some of the most energy-efficient foundation systems for passive solar homes, including the frost-protected shallow foundation.

Floor insulation is also important in other designs, for example, when thin-set tile is laid on wooden subflooring over crawl spaces or basements. As a general rule, thermal mass floors located over crawl spaces require more insulation than thermal mass floors over basements. The colder the climate, the greater the recommended level of slab and floor insulation.

**TABLE 3-1. Insulation Requirements by Climate**

| CLIMATE | INSULATION | |
| --- | --- | --- |
| | WALL | CEILING |
| Temperate | R-30 | R-60 |
| Cold | R-40 | R-80 |
| Hot | R-40 | R-80 |

Ken Olson and Joe Schwartz, "Home Sweet Solar Home, " *Home Power,* Issue 90 (August/Septembr, 2002), 86-94.

Air infiltration should be rigorously controlled as well. The tighter the home, the less energy it will require. However, bear in mind that making a home too airtight can render indoor air stagnant and even unhealthy. (I'll discuss this issue in chapter 6.)

## Designing Overhangs

Overhangs control the solar heating season, that is, the beginning and end of the period of solar gain through south-facing glazing. Fixed overhangs should be designed so there is a separation between the top of the window and the underside of the projecting surface, as shown in figure 3-17. This feature, combined with the length of the overhang, allows the low-angled winter sun (angle A in the drawing) to penetrate the interior, while blocking the high-angled summer sun (angle B) from gaining entrance at the end of the heating season. To determine the length of the overhang projection, use the following formula:

length of projection (L) = height of window opening (H)/F factor.

In this equation, the F factor is a number that varies with the latitude. It is determined from table 3-2. For example, suppose you are building a home in Wisconsin at 44° north latitude. Suppose your windows are 6 feet high. To determine the overhang, you would simply divide 6 feet by the F factor, which in this case is 2 to 2.7. As you can see, F factors are expressed in a range, which allows some design flexibility. If you want more sunlight, use the larger number in the range. Knowing your heating requirements (heating degree days) and solar availability (average daily solar radiation by season) will assist you in making this determination.

Generally speaking, the colder the climate, the greater the need for solar heat. The greater the need for solar heating, the smaller the roof overhang. But

**TABLE 3-2. F Factor Used to Determine Length of Overhang Projection in All Passive Solar Designs**

| NORTH LATITUDE | F FACTOR |
|---|---|
| 20° | 5.6 - 11.1 |
| 32° | 4.0 - 6.3 |
| 36° | 3.0 - 3.5 |
| 40° | 2.5 - 3.4 |
| 44° | 2.0 - 2.7 |
| 48° | 1.7 - 2.2 |
| 52° | 1.5 - 1.8 |
| 56° | 1.3 - 1.5 |

**FIGURE 3-17** (bottom left) *The fixed overhang regulates solar gain, determining when sunlight first enters a passive solar home and when it can no longer penetrate south-facing glass.*

**FIGURE 3-18** *Overhang protects walls and windows during the cooling season.*

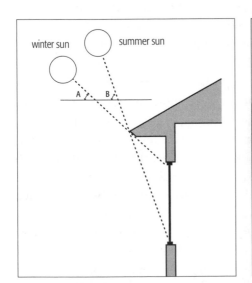

not all cold climates are created equal (figure 3-18). Some are cold and sunny. Others are cold and cloudy. Eastern Montana, for instance, falls into the first category. Maine and Vermont fall into the latter. A designer can adjust overhang in response to these differences. The less sunlight available, the shorter the overhang projection.

In most locations, a 2-foot overhand shades an 8- to 9-foot wall well, but there's more to designing overhangs than I've shown here. The distance between the top of the window and the soffit also plays a factor on controlling sunlight entry. Log on to the Web site www.susdesign.com/overhang/index.html for help designing overhangs for your region of the country.

## The Pros and Cons of Direct-Gain Passive Solar

Direct gain is the easiest-to-implement passive solar design strategy. It is also one of the easiest to botch. Early pioneers plunged into passive solar building with little understanding of the full requirements of the system—especially the need to balance mass and glass and the need to provide sun-free zones. Modern designs and modern design tools, such as the energy analysis software described in chapter 7, make the task of designing a highly functional, comfortable direct-gain passive solar home far easier.

Before you choose this approach, however, it is important to carefully consider the advantages and disadvantages of direct-gain passive solar design, in order to develop ways to compensate for its drawbacks.

On the plus side is the design simplicity. The formula is deceptively simple: Orient a house to the south, concentrate glass on the south wall, insulate well, and provide overhangs, shade, and thermal mass. The immediate benefit is a warm, comfortable home with extremely low heating and cooling requirements—so long as you've not made any major errors. Direct-gain passive solar homes "sail" lightly into the future with little impact on the environment, especially if you pay attention to other aspects of design, such as the size of the house and the materials used in making it.

Appearance is another advantage. Many passive solar homes are quite attractive (figure 3-20). Moreover, direct-gain passive solar design is amenable to a wide range of architectural styles. Passive solar homes fit nicely into almost any neighborhood. Furthermore, solar glazing can be located on the front, the back, or the side of the house. It is therefore suitable for homes located on either the north or south side of a road running east-west. As noted in chapter 1, direct-gain passive solar can also be incorporated into houses located along streets running north-south.

Another advantage of passive solar is light, airy interiors. Direct-gain homes are often bright and cheery. There is something almost magical about

**FIGURE 3-20**
*Passive solar can be quite attractive, as in this home in the Denver metropolitan area.*

One method for capturing solar energy in floors was devised by James Kachadorian, author of *The Passive Solar House*. This innovative designer and engineer, who has built many economical passive solar homes in the northeastern United States, creates thermal mass in the floors by installing concrete blocks under a concrete slab (figure 3-19).

The blocks are oriented in rows with the interior cavities pointing north and south to create continuous pathways for the flow of air. Over the top of this block matrix, a three- to six-inch slab is poured. Vents are created in the floor along the north and south sides of the house, so room air can circulate through the submass labyrinth.

In this design, sunlight enters the house through south-facing windows, then strikes the interior surfaces where it is converted to heat. Hot air flows by convection or, more commonly, with the aid of fans, into and through the subslab storage system, entering the vents at the back of the house, then re-emerging at the front (along the south wall). As it passes through the floor, this air gives off its heat, which is stored in the blocks. At night, the heat radiates from the blocks to the slab into the room, keeping the house warm year-round, even in the cold Vermont climate!

Although this technique seems to work well, especially if a fan is incorporated to blow air into the subslab labyrinth, some designers are concerned that moisture can build up in the passageways. The cool moist environment, they assert, could serve as a breeding ground for mold and mildew. Spores could become entrained in the air flowing through the system, contaminating indoor air and causing potentially serious health problems. Fans require a fair amount of energy, too.

Concerned about problems such as these, passive solar designer and builder Bruce Brownell of Adirondack Alternate Energy in Edinburg, New York, circulates warm air through a gridwork of pipes in the slab, a 70- to 100-ton mass storage system under the lowest floor. Warm air is delivered to the slab via air shafts that siphon heat from the ceilings. The pipes are easier to clean and less prone to mold and mildew than the cement block labyrinth. Like Kachadorian's, Brownell's houses remain warm and experience stable indoor temperatures despite extremely cold outside temperatures.

Michael Middleton

**FIGURE 3-19**

*The solar slab: Warm room air circulates beneath the concrete slab in pathways made from concrete block. Heat is absorbed by the blocks, then migrates into the overlying slab and is radiated into the room. Fans greatly assist in the movement of air into and through the solar slab. Some designers are concerned that the blocks may also foster mold growth, which could be hazardous to health.*

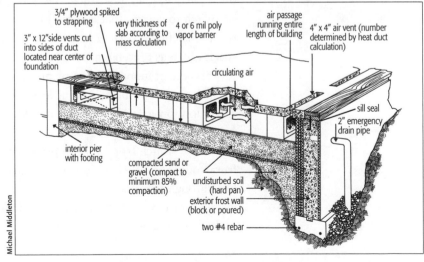

Michael Middleton

3/4" plywood spiked to strapping

3" x 12" side vents cut into sides of duct located near center of foundation

vary thickness of slab according to mass calculation

4 or 6 mil poly vapor barrier

air passage running entire length of building

4" x 4" air vent (number determined by heat duct calculation)

circulating air

interior pier with footing

compacted sand or gravel (compact to minimum 85% compaction)

undisturbed soil (hard pan)

exterior frost wall (block or poured)

two #4 rebar

sill seal

2" emergency drain pipe

solar light streaming into a home on a sunny winter day! My house comes alive when the sun shines in.

Direct-gain homes provide considerable daylighting, which lowers electric bills and can make for a much more pleasant working and living environment, provided you've designed ways to prevent glare and sun drenching, described in chapter 1.

Many solar-heated homes are built with an open floor plan to enhance heat distribution. This adds to their visual appeal. In addition, open floor plans allow for smaller, less resource-intensive houses.

South-facing windows provide views of one's surroundings. In the country, views of distant fields, forests, or mountain ranges are not uncommon. Views also greatly add to one's pleasure and to property values.

Because they are airtight, direct-gain passive solar homes do not suffer from drafts like many conventional homes. If the design includes sufficient, well-placed mass, direct-gain homes will be extremely warm and comfortable.

On the downside, there is much room for error. Overglazing and under-massing a home, common problems of the past, can cause severe overheating. Pay attention to the glass-to-mass ratio to achieve optimum comfort.

Direct-gain homes bathe occupants in light, which may cause considerable glare. Ultraviolet light can damage carpets, window coverings, and furniture, although newer glazings that limit UV penetration help reduce this problem. Create sun-free zones, as described in chapter 1, for comfort and optimum livability.

Direct-gain passive solar can also be hard on house plants, particularly tropical plants. During the summer, tropical plants may not receive enough light. During the winter, they may receive too much. Many tropical plants cannot tolerate direct sun. Intense sunlight streaming into south-facing windows can bleach and burn leaves. To keep from killing tropical plants an owner may need to move them out of the path of the sun in the heating season or switch to species that can tolerate bright sunlight. Cacti, succulents, more traditional flowers, such as geraniums, and hardy tropical and subtropical species such as dracena, hibiscus, bougainvillea, and orange trees, are suitable for this type of indoor environment.

Privacy is a problem in some locations, as large south-facing walls provide a window on a family's private world. Window shades can help solve this problem.

In the winter, large walls of glass become a frigid heat sink that robs heat from rooms and their occupants, greatly decreasing comfort. Excessive heat loss through windows can be a problem if shades or rigid insulation are not installed to cover windows at night. Opening and closing shades can be a bother to some people.

# INDIRECT-GAIN PASSIVE SOLAR HEATING

Indirect-gain passive solar is the most widely applicable passive solar design option, good for the mildest to the most severe climates. Indirect-gain systems are also known as thermal storage walls or Trombe walls.

Thermal storage or Trombe walls are mass walls located on the south side of houses. As shown in figure 3-21, clear glass is situated three to six inches away from a mass wall, typically made from poured concrete, concrete block, rammed earth, or other similar dense masonry or earthen materials.

During the winter, low-angled sunlight streaming through the glass strikes the dark-colored surface of the mass wall. Heat forming at the surface of the wall follows two routes. First, much of the heat is absorbed by the mass. This heat slowly begins to migrate into the wall, flowing from hot to cold. Second, the remainder of the heat is transferred into air in the space between the glass and the mass wall. It may be lost to the outside or captured by venting the wall, as described shortly.

Most thermal storage walls are designed so heat reaches the interior surface at about the time the sun sets in the winter, or slightly later. Thus, the solar-derived heat arrives at the inner surface of the wall just when heat is required. It then begins to radiate into the room and continues providing warmth and comfort throughout the night. The result is known as *delayed solar gain*.

The time lag, the time between the moment the sun begins to warm a thermal storage wall and the time the heat begins to radiate into a room, varies from a few hours to an entire day, depending on the thickness of the mass and

*Thermal storage walls are designed primarily to provide nighttime heating in climates that range from mild to severe. They are, therefore, the most widely applicable form of passive solar.*

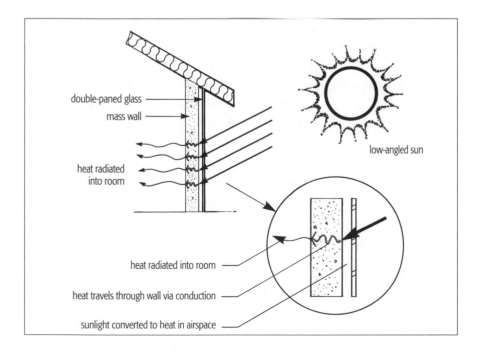

double-paned glass

mass wall

heat radiated into room

low-angled sun

heat radiated into room

heat travels through wall via conduction

sunlight converted to heat in airspace

**FIGURE 3-21**
*Cross section of a thermal storage wall or Trombe wall. Sunlight striking the dark surface of the thermal storage wall is converted to heat. In an unvented thermal storage wall, heat then migrates through the mass wall into the home.*

its density. Thermal storage walls may vary in thickness from six to twenty-four inches, though eight to eighteen inches is most common. How thick the wall should be in a home depends on the masonry material and the time lag required. Table 3-3 shows the recommended thickness by material. The thicker the wall, the longer it will take the heat to reach a room. In addition, the thicker the wall, the less daily variation will occur in the surface temperature of the inside wall surface.

Some designers substitute water for masonry materials in thermal storage walls. Water has a higher heat capacity than masonry, meaning it holds more heat per unit volume. However, water releases heat more quickly.

In most cases, free-standing plastic water columns are situated behind the glass to absorb sunlight and store heat. At night or during cold, cloudy periods, the mass walls radiate heat into the adjacent room. In most installations, water tubes are visible to the occupants of the room to optimize their performance. Applying wall board may enhance the aesthetics, but it decreases heat transfer from the water to the room air.

Windows can be installed in thermal storage walls to permit an outside view, daylighting, direct-gain solar heating, and an escape route in case of fire. (Building codes typically will require openable windows for emergency egress in all bedrooms.)

Trombe walls may also contain air vents to siphon heat out of the space between the glass and the mass wall, providing immediate or daytime solar heat gain. Here's how they work: As shown in figure 3-22, cool room air enters the air space through the lower vent where it is warmed by sunlight. As it flows upward, it gains additional warmth, creating a convection current that propels the air through the top vent into the room while drawing cool air in through the bottom vent. Air enters the room at about 90°F.

Vents need to be sized correctly to ensure adequate air flow. As a general rule, for every 100 square feet of Trombe wall, you will need 2 square feet of vent, divided equally between the top and bottom vents. A 300-square-foot wall will therefore require about six square feet of vent, three square feet on the bottom and three square feet on the top.

When thermal storage walls are vented, it is almost always a wise decision to install closable top vents to prevent heat loss at night. Nighttime heat loss occurs as a result of a reverse convective loop that siphons warm air out of the house, beginning when the sun no longer shines on the wall. As the convective loop that heated the room falters, the reverse convective loop begins to operate. As shown in figure 3-23, warm room air flows into the air space through the top vent. Inside the air space, the warm air cools

### TABLE 3-3. Mass Wall: Materials vs. Thickness

| MATERIAL | DENSITY (lb/cf) | THICKNESS (inches) |
|---|---|---|
| Concrete | 140 | 8 - 24 |
| Concrete masonry (concrete blocks) | 130 | 7 - 8 |
| Clay brick | 120 | 7 - 16 |
| Light Weight Concrete Masonry | 110 | 6 - 12 |
| Adobe | 100 | 6 - 12 |

Source: *Passive Solar Design Strategies: Guidelines for Home Building,* Sustainable Buildings Industry Council, National Renewable Energy Laboratory, and Charles Eley Associates.

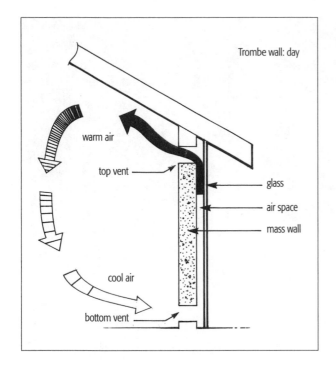

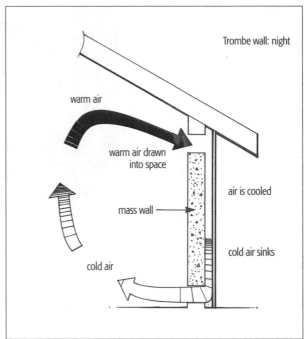

and sinks, sucking more warm air in through the top vent and pushing cold air in through the bottom vent.

The reverse convection loop draws heat out of a house, subtracting from heat gain unless backdraft dampers or operable louvers are installed in the vents. Both are effective deterrents, but do require daily operator involvement. That is, they must be opened and closed each morning and night. (In my previous home, I found it easier to block the vents entirely and rely on delayed gain and immediate gain from a window in the thermal storage wall.)

As in all other successful passive solar features, mass walls must be designed to prevent overheating during the cooling season. Overhangs are the most common form of protection. Use the equation presented earlier in the chapter to determine the length of the overhangs. Shading may also be provided by movable insulation panels or other devices, which are situated over the glass during the day. For a list of options, see the sidebar on this page.

Thermal storage wall systems may also require protection in the winter, to prevent nighttime heat loss, especially in cold climates. To solve this problem, some builders rely on rigid foam insulation that is placed over the glass at night. We'll examine various options below.

Although thermal storage walls are efficient, many individuals avoid them, fearing they will be ugly. The truth is, from the outside a thermal storage wall looks like ordinary south-facing glass, unless you are standing very close to it (figure 3-24). Inside, thermal storage walls can be finished with plaster or stucco and thus can be made quite attractive. Windows in the wall open it up to views, so occupants don't feel as if they are imprisoned by massive concrete walls.

**FIGURE 3-22 (left)**
*In a vented thermal storage wall, heat may also be transferred to the adjoining living space by natural convection.*

**FIGURE 3-23 (right)**
*At night, heat can be drawn out of a house by a vented thermal storage wall, if the vents are not closed off. Notice the reverse convection loop.*

**BUILDING NOTE**

From a distance, a Trombe wall looks just like ordinary glass. Windows can be built into the wall, too, so a room need not be closed off from sunlight.

FIGURE 3-24

*From a distance, a thermal storage wall in the author's previous home looks like ordinary solar glazing. The window in this thermal storage wall permits some direct gain and a view of the outside world.*

## Keys to Successful Design of Indirect-Gain Systems

Thermal storage mass walls are ideal solar collectors, but as in other solar designs there is room for error.

*First*, be sure the thermal storage wall is oriented due south plus or minus 10 degrees. As in any passive solar system, the further you deviate from true south, the lower the winter heat gain and the greater the summer heat gain.

*Second*, high-performance glass is recommended for use in thermal storage walls, although some builders have successfully used fiberglass, acrylics, and polycarbonates (Plexiglass). Remember: The more opaque the glazing, the lower the solar gain.

As a rule, double-pane glass is best. The colder the climate, the more essential it is to choose a glass with a high R-value (or low-U value). High-performance glass includes products made with low-e coatings (thin films on the glass itself), suspended plastic films (for example, Heat Mirror glass), and gas fills (argon gas between the panes). Single-pane glass is generally not advisable, except in the mildest climates. In such instances, movable insulation may be required to optimize the system's performance.

*Third*, be sure that the glass facing is thermally isolated from the mass wall. If it isn't, heat may escape from the thermal storage wall via conduction at night or during cloudy periods. Metal framing is a source of thermal bridging and will conduct heat out of the wall at night, if it is not isolated from the mass. Wood framing is a better option; however, wood takes a beating in the intense (150 to 180°F) heat in the airspace between the glass and the mass wall.

*Fourth*, caulk and sealants used to construct thermal storage walls should be able to withstand considerable expansion and contraction resulting from daily temperature fluctuations. Don't skimp. Buy the best.

*Fifth*, paint applied to the surface of a thermal storage wall should be capable of withstanding high temperatures. Better yet, you may want to install a selective surface material to the exterior surface of the mass wall. Selective surface material usually comes in thin sheets that adhere to the wall. Because the dark solar surface of the material absorbs virtually all of the sunlight that strikes it, very little heat is released into the airspace, from which it can escape through the glass. The result: greater solar gain. The colder the climate, the more important this material becomes. (Note that the increase in efficiency created by the selective surface reduces the need for nighttime insulation.)

*Sixth*, thermal mass walls must be dense and capable of absorbing a significant amount of heat and giving it off slowly. Masonry materials such as concrete and brick are ideal. Concrete block filled with sand or concrete work well, too. To cut down on labor, some builders dry stack concrete blocks, rather than mortar them. They then use a surface-bonding material to hold the block wall together. Surface bonding combined with dry stacking saves considerably on labor costs. This is an easy process to learn, and is well suited for less-experienced owner-builders.

Earthen materials such as rammed earth and adobe work well, too, but they're less dense than the previously mentioned materials. Whatever you use, be sure to check with local building codes for any structural reinforcement requirements for mass walls. Remember, these walls generally also serve as load-bearing exterior walls.

*Seventh*, the interior surface of a mass wall should be minimally finished to enhance heat transfer. Earthen, gypsum, and lime plaster work well, as do cement or synthetic stucco. Drywall will work, so long as it is in direct contact with the mass wall. Furring drywall against a mass wall is a bad idea because it reduces heat conduction. (Note that all of these products, except earthen and lime plaster, outgas potentially toxic chemicals into the room; lime plaster is caustic and dangerous to work with.)

*Eighth*, remember that insulation is generally needed most in colder climates. As engineer Al Eggen, of K.T. Lear Associates in Ashford, Connecticut, notes, thermal storage walls can be a "disaster in climates where the sun does not shine every day." He goes on to say, "Unless you can wrap an R-20 or so blanket over the outside wall, a few cold, cloudy days results in a very large cold wall." To avoid this problem, designers rely on two basic options: (1) external insulation positioned against the exterior surface of the glass at night, for example, rigid foam insulation panels; and (2) internal insulation, that is, devices located in the airspace between the glass and the mass wall. Both internal and external insulation work well, although the latter is more difficult to install and service. Rigid foam exterior insulation generally ranges from R-4 to R-8 per inch. If you are using rigid foam, it should be at least the same size as the glazing and should fit tightly over the glass at night. If you choose interior

## SELECTIVE SURFACE FOILS AND PAINTS ON THERMAL STORAGE WALLS

Some builders apply a special dark-colored paint or adhesive foil known as a *selective surface foil* to the exterior (sun-facing) side of thermal storage walls. Selective surfaces absorb solar radiation just like any black surface, but reduce the infrared emittance (heat radiation), thereby greatly reducing heat loss from the exterior surface of the mass wall. A high-quality selective surface performs similar to R-9 moveable insulation, but requires no operator involvement. However, because selective surfaces may not be aesthetically appealing, it is best to hide them with high-transmittance translucent glass instead of clear glass.

insulation, you will need a more sophisticated system. Be sure that internal insulation does not interfere with the workings of the thermal storage wall.

## Region-Specific Design of Thermal Storage Walls

The design features of a thermal storage wall, like other systems, depend in large part on the heating requirements and available sunlight. Sunny and mild climates are less demanding than cloudy and cold ones.

*Mild Climates.* In mild climates, a thermal storage wall should work well with single-pane glass, provided it has sufficient overhang to protect it from off-season heating. The thermal storage wall, however, serves primarily as a nighttime heat source via delayed solar gain. If daytime heating is desired, you may want to install a window or two in the wall (for direct gain) or vents to draw warm air into the room, as described above.

*Moderate Climates.* As winter heating requirements escalate, some modifications can be made to a house and a thermal storage wall system to optimize their performance. Closer orientation to true south increases heat output, as does double-pane glass with a high R-value (low U-value). As in mild climates, installing windows in a thermal storage wall or elsewhere on the south wall provides more immediate solar gain. Installing vents in the wall also provides daytime heat. However, as with many other design features, there's a tradeoff: venting decreases the amount of heat that is available at night. When asked about his preference, Ron Judkoff of the Natural Renewable Energy Laboratory emphatically sided with the direct-gain approach. His advice: "Design non-circulating (unvented) thermal storage walls instead of circulating (vented) TSWs. Install windows for direct gain."

In cold climates, insulation may also be required to reduce heat loss from the wall at night or during cloudy periods.

*Moderate-Severe to Severe Climates.* In still colder climates, thermal storage walls may require further modifications: even closer orientation to true south, low-U-value glass, installation of windows for immediate gain, and use of circulating fans. Selective surfaces create better thermal transfer. Builders should pay even more careful attention to reducing conduction losses and sealing air leaks, and care should be taken when applying interior finishes on the wall to permit maximum heat transfer into adjacent rooms.

External insulation to cover the glass at night is also highly desirable. Further gains can be made if the insulation panels are reflective. Insulated reflector panels may lie on the ground during the day or can rest on a wooden support structure. They should be tilted about 5° away from the house to permit water or melting snow to drain away from the foundation. When opened during the day, insulated reflector panels direct more light onto the thermal storage wall and can increase the heat gain by up to 30 or 40 percent. Reflective panels increase solar gain in situations in which thermal storage walls are shaded by trees. For those dedicated to achieving maximum self-sufficiency, they are highly recommended.

In more severe climates, designers can also increase the size (dimensions) of the thermal storage wall. As a general rule, the farther north the building is, or the colder the climate, the larger the thermal storage system should be. Table 3-4 provides helpful sizing information. To see how this table is used, let's assume you live at 48° north latitude and that the average outdoor temperature in January is 30°F. Go to the far right column, find 48° north latitude, then drop down to the third row. This number 0.70 is the ratio of masonry wall to floor area—that is, how large a mass wall should be in relation to the heated floor space. If the room is 400 square feet, you'll need 280 square feet of thermal storage wall (400 × 0.7 = 280). If average temperature and latitude are not shown on the table, you can interpolate—that is, find the value between measurements given on the table or use the closest one you can find.

## The Pros and Cons of Thermal Storage Walls

Like other passive solar design elements, thermal storage walls have their advantages and disadvantages.

On the positive side, thermal storage walls have the widest applicability. They can be used in mild to severe climates. They can even be used to passively cool homes, as explained in chapter 6. Secondly, thermal storage walls greatly reduce sun drenching, thereby lessening glare and damage to carpets, upholstery, and plants, which can be quite significant in direct-gain designs. Consequently, thermal storage walls are great for home offices, and reading or television rooms. I also recommend them for bedrooms, especially for those looking for a design that ensures a dark room at night.

*Early in the development of thermal storage walls, reverse air flow in thermal storage walls caused many people to criticize them as inadequate. But as noted in the text, the problem can be easily remedied by installing backdraft dampers or operable louvers.*

### TABLE 3-4. Sizing a Thermal Storage Wall

| AVERAGE WINTER OUTDOOR TEMPERATURE (Clear day)* | RATIO OF MASONRY WALL TO FLOOR AREA | | | |
| --- | --- | --- | --- | --- |
| | 36°NL | 40°NL | 44°NL | 48°NL |
| Cold Climates | | | | |
| 20°F | 0.71 | 0.75 | 0.85 | 0.98 (with night insulation) |
| 25°F | 0.59 | 0.63 | 0.75 | 0.84 (with night insulation) |
| 30°F | 0.50 | 0.53 | 0.60 | 0.70 |
| Temperate Climates | | | | |
| 35°F | 0.40 | 0.43 | 0.50 | 0.70 |
| 40°F | 0.32 | 0.35 | 0.40 | 0.44 |
| 45°F | 0.25 | 0.26 | 0.30 | 0.33 |

* Temperatures listed are for December and January, usually the coldest months.

Source : Steven Winter Associates, *The Passive Solar Design and Construction Handbook*

Thermal storage walls work best when nighttime heating is the primary goal. However, as noted above, they are also quite adaptable. By installing vents and windows for direct gain they can be modified to contribute to daytime heat demand as well.

Another advantage of thermal storage walls is that they provide mass in a relatively concentrated area, taking up a minimal amount of living space. And they provide great comfort. Rooms tend to be thermally stable and quiet. Finally, thermal storage walls are aesthetically appealing, externally and internally.

On the downside, thermal storage walls may add to the cost of construction, primarily by increasing the size of the foundation required to support the additional mass.

Thermal storage walls reduce daylight and access to views. Heat loss can be quite significant at night unless the exterior surface of the structure is insulated. Covering a wall each night and closing off the vents require additional effort on the part of the homeowner. Placing rigid external insulation over the glass, for example, requires the homeowner to venture outside on cold winter nights.

If you want to learn additional technical details about thermal storage walls, I recommend *The Passive Solar Design and Construction Handbook* by Steven Winter Associates. It provides numerous sketches of wall sections, showing various construction details.

## ATTACHED SUNSPACES: ISOLATED-GAIN SOLAR SYSTEMS

Isolated-gain solar designs, or more commonly, attached sunspaces, are used in a wide variety of climates, ranging from moderate to severe. As its name implies, an attached sunspace is a passive solar structure attached to the south side of a house (figure 3-25). Attached sunspaces can be used alone or in conjunction with direct-gain and indirect-gain systems and often do double duty—that is, they produce heat and provide additional living space, although careful attention to detail is required to obtain this sometimes elusive benefit. In other instances, they are designed to gain heat and serve as growing areas, hence the name solar greenhouses or attached greenhouses.

In an attached sunspace, sunlight penetrating south-facing and (in some cases) roof glass is converted to heat after being absorbed by interior surfaces.

FIGURE 3-25
*Attached sunspaces are an attractive solar design feature and are ideal for solar retrofits, but they are also fraught with problems. Be careful.*

Some of the heat warms the sunspace. The rest is transferred into the house. Getting the balance correct can be difficult.

FIGURE 3-26
*Projected and enveloped sunspaces.*

## Basic Options

Attached sunspaces come in two basic varieties: those that project from the south side of a house, and those that are enveloped by the south wall, as shown in figure 3-26. In the first instance, the sunspace "shares" one side with the house. In the second, it "shares" three sides.

Of the two approaches, the most common is the "projecting design." Although these are more popular and are ideally suited for retrofitting homes, they are more exposed to the elements. As a result, they lose heat more readily at night. In addition, homeowners find that it is often more difficult to transfer heat into a house from a projecting sunspace than from an enveloped sunspace. As a result, this type of sunspace is more likely to overheat.

Attached sunspaces may be located at ground level, give or take a little, or below the ground-floor level (figure 3-27). Ground-level sunspaces require good perimeter and subslab insulation to guard against heat loss. Below-ground-level sunspaces are sometimes referred to as pit designs. They tend to be slightly more thermally stable, as they are somewhat sheltered from temperature changes by the earth. In cold weather, the partial subterranean design may even result in some heat gain from the ground.

*Attached sunspaces enveloped by the south wall suffer less heat loss than a projecting design. They also transfer heat more readily to a house.*

FIGURE 3-27
*Ground level (a)and subterranean (b) attached sunspaces.*

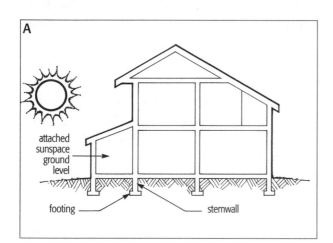

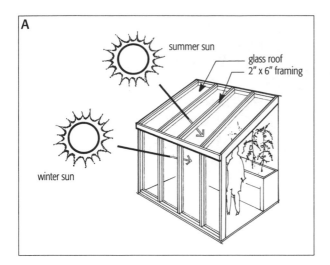 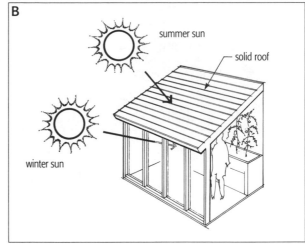

FIGURE 3-28
*All-glass (a) and south-facing-glass-only (b) attached sunspaces.*

*As a rule, all-glass sunspaces are less likely to provide useful living space than designs with only south-wall glass; the former are either too hot and sunny during the summer or too cold in the winter, unless they're well shaded and well insulated during these times.*

Attached sunspaces may be "all glass"—that is, may contain glazing on the roof and the walls—or may be designed with only south-wall glass (figure 3-28). Generally, the more glass, the more dramatic the temperature swings, and the less useful the space will be for humans and plants. In addition, all-glass designs suffer from intense sun drenching.

A far more useful approach is an attached sunspace with south-wall glass only (figure 3-28b). Solid roof design and overhangs permit less sunlight to enter the sunspace but insulation in the roof helps retain heat, making the structure more efficient than an all-glass design.

Sunspaces can be an extension of living space or can be thermally isolated from them (figure 3-29). In other words, they can open directly into a living space or be separated from living space by a wall. The first is known as an open-wall design; the second is a common-wall design.

***Open-Wall Design.*** In the open-wall design, the attached sunspace opens directly onto adjacent rooms—for instance, a living room—and is therefore considered an extension of the living space (figure 3-29a). Because this design opens into living space, the temperature of the sunspace is generally maintained within the thermal comfort range of the rest of the house. To do this, heat must be supplied to the sunspace during the evening and during prolonged cloudy periods. The colder the outside temperature, the greater the heat demand. In the summer, the area may overheat and tax a home's natural ability to cool itself or place additional demands on mechanical cooling systems. The more glass the unit has, the greater the risk that an attached sunspace will rob heat from a home in the winter and cause unwanted heat gain in the summer.

To minimize heat loss during the winter and heat gain during the summer, open-walled sunspaces must be well insulated. High-performance glazing, insulated shades, perimeter insulation, and subslab insulation are all vital. The more extreme the climate, the more drastic the measures that need to taken.

*Common-Wall Design.* The common wall sunspace design isolates the sunspace from the living space with a wall. There are three variations on this theme.

The first common-wall sunspace option involves use of a glass wall (figure 3-29b). In gentler climates, single-pane glass works well. In more severe climates, double-pane glass is advisable.

In the glass-wall design, solar energy streaming through the south-facing glass strikes solid surfaces within the sunspace and is converted to heat, thus warming the room air. Heat generated inside the sunspace is allowed to enter the home, typically through a sliding glass or patio door that connects the sunspace and living area. If the sunspace is not deep, and most are not, some sunlight may also enter the adjacent living space, providing direct gain.

In this design the attached sunspace is treated as an isolated space, not an integral part of the living space. The sunspace is therefore not generally heated or cooled. When the sun sets at night and the heat flow from the attached sunspace ceases, the collector space begins to cool down and is then closed off. The next day, the cycle repeats itself. To increase efficiency and comfort, removable insulation or insulated shades can be used to reduce heat loss through the glass wall or the south-facing glass.

The next permutation of the common-wall design involves the construction of a standard wall, usually a framed wall, between the sunspace and the adjacent room. The wall may be insulated or not, depending on the climate. In

**FIGURE 3-29**
*Attached sunspaces. (a) Open and three common-wall designs: (b) Glass wall. (c) Standard wall. (d) Mass wall.*

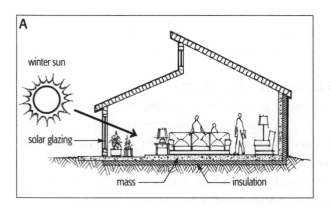

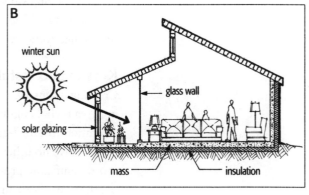

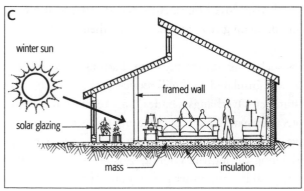

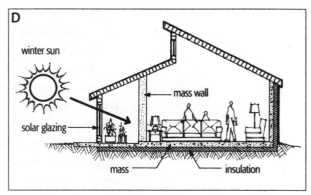

Open wall—no divider between sun-space and living space.

Common wall—common wall sepa-rates sunspace from living space. There are three variations on this theme: glass wall, standard wall, and high-mass wall.

more severe climates, insulation becomes much more important in preventing heat loss at night or during cloudy periods. In milder climates, insulation is generally not required during the heating season but does help reduce unwanted heat gain during the heating season. The Sustainable Buildings Industry Council recommends insulating a common wall to about R-10.

As in other designs, shades may be required to prevent overheating of the sunspace during the summer and early fall. Some designers install vents in the exterior wall of a sunspace to permit outdoor air to enter during the summer to help maintain a reasonable interior temperature. (This topic will be dis-cussed in chapter 5)

As illustrated in figure 3-29c, vents in the common wall can be used to transfer hot air from a sunspace to the living space. However, doors and win-dows are more commonly used. As a general rule, doors work better than win-dows. If a doorway is used, the opening should be at least 15 percent of the sunspace glass area. Because they are less efficient at transferring warm air, windows should comprise about 40 percent of the entire common wall area.

The final variation on the common-wall theme involves the use of a high-mass wall between the sunspace and living area (figure 3-29d). In this design, sunlight striking solid surfaces in the sunspace, including its mass wall, is con-verted to heat. Heat warms room air and then flows into adjacent living areas through vents or doors and windows in the wall, providing daytime heat. As in a thermal storage wall, heat absorbed by the mass wall also migrates inward, moving slowly toward the living area and providing significant night-time heating. Window shades and removable insulation can be installed to reduce nighttime heat loss or heat loss during cold spells.

## Keys to Successful Design of a Sunspace

Creating a sunspace that heats a house while providing living space or a grow-ing area is challenging—so challenging that I advise readers to give strong con-sideration to direct-gain systems. If your heart is set on an attached sunspace, or you have no other option, proceed with caution. Remember that one of the most significant problems with isolated-gain systems is their propensity for overheating, both in summer and winter. Overheating makes it difficult to grow plants or hang out with your loved ones. Attached sunspaces also typi-cally suffer from intense sun drenching year round, making them less than use-ful living space.

Because overheating is such a significant problem, I strongly recommend against all-glass designs. A well-insulated roof will reduce heat loss in the win-ter and heat gain in the summer. Overhangs, shades, and vegetation (such as deciduous trees) also protect sunspaces from overheating and sun drenching, but there is almost always a price to pay. For instance, roofs and overhangs required to prevent overheating during the cooling season make it next to impossible to grow fruits, vegetables, and other plants year-round. In the heating season, a roof permits greater personal use of a sunspace, but the trade-off is reduced heat gain.

Attached sunspaces may be built on site or purchased as prefabricated kits. Both options have advantages and disadvantages. Site-fabricated sunspaces can be designed to fit the home and therefore may blend better with the architectural style. However, on-site construction requires considerable know-how. For homeowners, be sure that your architect and builder understand the full requirements of a sunspace, for example, the need to protect all elements (such as framing members) from intense sunlight and moisture. Metal cladding helps to protect exterior surfaces of wooden framing members from sunlight, rain, and snow on the outside and humidity and high heat on the inside.

Because site-fabricated sunspaces are designed in conjunction with the house, adequate protection in the form of overhang, shading, and insulation is generally possible. However, despite these advantages, homemade sunspaces are notorious for leakage. This occurs when seals between glass and the framing members fail due to excessive expansion and contraction resulting from changes in daily temperature. Use the best-quality materials. Kiln-dried lumber is better than rough-sawn or green wood. It warps and cracks less. No matter what wood you use, be certain that it is protected from interior moisture. Painting inside surfaces a light color helps reduce heat absorption by framing members. Use high-quality glass rather than plastic for your glazing material, too. Plastics can discolor, cloud over, and deteriorate over time. And be sure to install a good system for insulating and shading the glass. Leave room for a thermostatically controlled fan system in case you need it.

Prefabricated models offer fewer options than site-fabricated sunspaces and may not always be visually compatible. In addition, most prefabricated models are all glass. Some difficulty may be encountered when installing a prefabricated unit, if the house has not been precisely designed to fit the structure.

Although prefabricated sunspaces are generally pretty expensive, they are also generally quite weatherproof. They are usually made from anodized metal and require little maintenance. However, they may be difficult to insulate. They will require the construction of a foundation and mass floor. Sturdy and weather resistant as they are, metal frames can result in significant heat loss, unless thermal breaks are provided.

As in direct-gain systems, south-facing glazing in an attached sunspace can be either vertical or angled (tilted). Vertical is probably the best bet because it is less likely to overheat, is easier to shade, and less prone to water damage and leakage.

## Building an Attached Sunspace for Your Climate

Sizing an attached sunspace can be tricky. In most instances, the footprint dimensions are determined first, usually to create adequate living space or sufficient growing area. Solar glazing is then determined.

Requirements for south-facing glass vary by climate. The colder the climate, the more window space is advisable. Table 3-5 lists the square feet of glazing for each square foot of floor area by average outdoor temperature. The SBIC recommends three square feet of 4-inch thermal mass for every square foot of solar glazing. Mass can be in the floor, walls, and freestanding—for example, planters or water tubes or water barrels. In the open-wall design, thermal mass inside the living space can also be used to absorb heat and sunlight.

Mass common walls should be four to eight inches thick, depending on the lag time. Interior vent openings should be no more than 3 percent of the total area of the common wall separating the sunspace from the living space.

As a rule, the colder the climate the more important it is to insulate the foundation, slab, and glass. Insulation is extremely important in dual-function attached sunspaces—those that provide heat but also serve as living spaces or growing areas.

If the sunspace is thermally isolated from the living area and is just used as a giant heat collector, perimeter insulation is required but subslab insulation is not. That said, the system will perform better and will be less subject to wide temperature swings if you insulate under the slab. Insulation is inexpensive, so it makes sense to take measures to optimize the performance of the system. I built an attached sunspace on my previous home, insulated under the floor, and monitored temperatures over the winter. Even though temperatures frequently dropped below zero outside and the sunspace glass was not insulated, greenhouse temperatures only dropped below freezing two times during our brutal winter, and one of those times I accidentally had left the sliding glass door open a little.

Shading is usually best accomplished by devices on the outside of the glass, especially for roof glazing. Greenhouse suppliers and others offer a durable shading fabric. Finally, overhangs are determined the same way as in other types of passive solar systems.

In either case, east- and west-facing walls should probably be opaque (solid) and well insulated. They won't gain much heat during the heating season, and could contribute to overheating in the cooling season. Small windows, however, may need to be installed in them for cross ventilation and cooling.

## Pros and Cons of Isolated Gain

Attached sunspaces are very popular, but they do pose some significant problems. What are the advantages and disadvantages of this technique?

Attached sunspaces provide heat for themselves and adjacent rooms. They may also reduce heat loss

### TABLE 3-5. Glazing Requirements of an Attached Sunspace

| AVERAGE OUTDOOR TEMPERATURE in December and January on a Clear Day | SQUARE FEET OF SOLAR GLAZING per Square Foot of Floor Area |
|---|---|
| Cold Climates | |
| 20°F | 0.90 - 1.50 |
| 25°F | 0.78 - 1.30 |
| 30°F | 0.65 - 1.17 |
| Temperate Climates | |
| 35°F | 0.53 - 0.90 |
| 40°F | 0.42 - 0.69 |
| 45°F | 0.33 - 0.53 |

Source : Steven Winter Associates, *The Passive Solar Design and Construction Handbook*

from the main living area. Even on cloudy days, sunspaces provide a degree of thermal protection much like mud rooms or airlocks. In addition, certain designs—those with solid walls separating the sunspace from the house—reduce glare and sunlight penetration into the living space. As a result, they're easier on occupants, furniture, and furnishings than direct-gain systems.

With careful design, attached sunspaces can provide additional living space and can be used to grow plants year-round. An attached sunspace also adds beauty to a home.

Attached sunspaces can also provide warm air that can be circulated though the exterior walls of a home, as shown in figure 3-30. This design is known as an *envelope house* and is discussed in more detail in the accompanying box.

The main disadvantage of attached sunspaces is that they frequently fail to measure up to expectations. In many cases, sunspaces provide their own heat, but not enough to be of great benefit to the rest of the house.

Another significant problem is that it is often difficult to transfer heat generated inside a sunspace to neighboring rooms. Heat doesn't flow laterally very well. Therefore, fans may be required to move warm air from a sunspace into adjacent living space.

Sunspaces also tend to overheat. In the winter, they are often exceedingly hot and drenched in blinding sunlight. Both problems render them useless as daytime living space. Moreover, at night sunspaces are often too cold for comfort. Thus, a sunspace may add some heat, but provide no extra useful living space. In essence, it becomes a huge and costly solar collector. The investment in living space of dubious merit to gain a small amount of heat makes this system extremely costly. You would most likely be better off with a direct- or indirect-gain design. They provide positive, fairly predictable heat gains with little additional cost or material investment.

Sunspaces often make lousy growing areas as well. As noted previously, all-glass attached sunspaces tend to overheat in all but the coldest months, especially if glass is incorporated in the roof design. Most plants can't take heat over 85°F. Sunspaces with south-wall glass only are also pretty useless as a year-round greenhouse. Although they may have plenty of light for plants during the heating season when the sun is low in the sky, plants become leggy or fail to thrive during the summer when the sun is high in the sky and unable to illuminate the growing area.

## COMBINING SOLAR DESIGN FEATURES

Greater solar gains, as required in colder climates, may be possible by combining solar designs, that is, including direct gain and thermal storage walls. In my house, the vast majority of the solar design involves direct gain. However, airlocks for the two doors leading to the outside are enveloped sunspaces. They

### LIVING SPACE OR GROWING AREA?

For sunspaces designed to provide heat and living space, units with little or minimal roof glazing are your best choice. Insulate the roof as you would a house roof. Be advised, however, that it is very unlikely that there will be enough overhead light in the summer to grow vegetables or other sun-loving plants. If you want to grow food in this space, you may need to use a grow light in the summer. A small photovoltaic system designed to power the grow lights will cut down on electrical demand, but the cost may be prohibitive for the tiny amount of electricity you're going to use. If your utility offers a choice of green power, for example, wind power (many do), that might be a better option

For year-round or near year-round cultivation, you'll need some roof glass—either skylights, full glass, or suntubes. But in such instances, overheating is very likely going to occur—not just in the summer, but very possibly in the winter as well.

Over the years, many people have been attracted to the concept of the envelope house, a passive solar design that circulates heat generated by an attached sunspace through a continuous cavity in the building envelope. This heat radiates into the interior space, warming the occupants.

Also known as a *thermal envelope house* or a *double envelope house,* this "house-within-a-house design" requires construction of a double building envelope with an airspace, measuring six to twelve inches. During the day, sunlight entering the sunspace on the south side of the house is converted to heat. Heat from the attached sun space rises and flows by convection through the continuous airspace.

As shown in figure 3-30, air travels from the sunspace through the roof. It then travels down the north-wall airspace and through the floor cavity, reentering the sunspace through openings in the floor. This completes the convec-

tive loop. As it flows along its convective loop, heat radiates into the house.

During the winter, occupants of a thermal envelope house are surrounded by a blanket of warmth. Some heat may be stored in the floor (for example, a concrete slab over the air space) or some other form of mass, such as a bed of gravel or the earth itself. At night, stored heat radiates into the living space.

The thermal envelope house was a popular design theory in the late 1970s, but few of them were built. One reason was that construction costs were much higher than a standard direct-gain passive solar home. Brookhaven National Laboratory studied one envelope house during the winter of 1980 and found that the house performed fairly well; for example, it had lower energy bills than a comparable solar home, largely due to the high thermal insulation value of the double-wall construc-

tion. However, researchers found that heat was not evenly distributed. They also found that the envelope house performed no better than a superinsulated house, which is cheaper to build.

Another drawback of the thermal envelope house is the potential for catastrophic fire. The interconnected system of airspaces in the shell of the home provides an avenue through which flames can easily spread. The structurally protected chase is difficult to access in case of a fire. As a result, anyone seriously considering building a thermal envelope house should install a fire sprinkler system.

National Renewable Energy Laboratory's Ron Judkoff commented, "Don't build envelope houses. They are wasteful of space, material, and money, and often don't work very well." Storing heat under a floor can be a dangerous proposition, as well. If moisture is present in the air, it can condense in the subfloor storage area, creating ripe conditions for mold and mildew. Spores can circulate in the air, causing health problems.

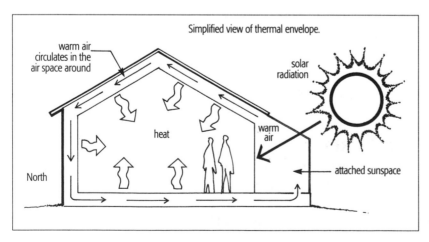

Simplified view of thermal envelope.

warm air circulates in the air space around

solar radiation

heat

warm air

North

attached sunspace

FIGURE 3-30
*The thermal envelope house, a passive solar design. Warm air from the attached sunspace circulates through the walls, roof, and floor, providing a layer of warmth surrounding the living space. Although they perform well, these homes cost more and have other problems that make them impractical.*

serve three functions: they prevent cold air from entering the house when people come and go, they gain heat, and they are a repository for coats and shoes.

In passive solar homes with two or more solar features, the total allotment of south-facing glass can be increased substantially, but as noted earlier should not exceed 20 percent of the floor space. In a 2,000-square-foot home, for example, 400 square feet of solar glazing is permissible. A designer could include 240 square feet of direct-gain glazing (12 percent of the floor space) and 160 square feet of sunspace glass (8 percent of the floor space). Remember: Additional mass will be required to accommodate the solar gain from this additional glass.

## ART, SCIENCE, AND ARCHITECTURE

Designing a passive solar home requires a melding of art, architecture, engineering, and science. Art and architecture determine the design of a passive solar home. To fine-tune a design, engineering and science come into play. Their role is to ensure that the design works well, structurally and thermally, in the intended site. The four areas blend to produce a structurally sound, aesthetically appealing home that provides economical comfort year-round.

The rules given in this chapter provide the general guidelines for passive solar design. Remember to think holistically: An integrated approach is vital to success. To be certain the design you've arrived at will perform well, you or your architect need to perform calculations to determine how a home will function given the temperature, solar availability, and house design. In chapter 7 I'll introduce you to two methods of running those numbers: one old-fashioned approach and one involving sophisticated but user-friendly computer software programs that perform the long, arduous calculations in the blink of an eye.

# SUPPLYING BACK-UP HEAT SUSTAINABLY

ALMOST ALL PASSIVE SOLAR homes require some form of back-up heat. Even those designed to achieve 100 percent of their heat from sunlight may require a back-up heat source from time to time, for instance during unusually cold and cloudy periods.

Ensuring year-round comfort is one of the main reasons for installing a back-up heating system in a passive solar home. But back-up heating systems are also required by most local building departments. In addition, back-up heat may be required to secure a mortgage, as well as to sell a home. Buyers who are unfamiliar with the tremendous potential of passive solar heating may be wary of a homeowner's claims that his or her house actually "heats itself with sunlight" and may insist on some form of supplemental heat. If your house does not have a reliable back-up heating system, buyers may select another home.

## OPTIONS AND CONSIDERATIONS

Selecting a back-up heating system is no easy task. There are four types of systems: forced air, radiant floor and baseboard, wall-mounted heaters, and wood stoves and masonry heaters (see sidebar). However, within some categories, such as radiant heating systems, you have several choices of fuel and different heat sources. Forced-air and radiant-floor systems, for instance, can be supplied with heat from oil- or gas-burning furnaces or boilers, heat pumps, wood furnaces, or solar hot water systems.

The system you select depends on many factors (see sidebar on page 130). For many people, cost and heating capacity—that is, the amount of heat a system produces—are the major criteria for selecting a system. When building an energy-efficient, passively conditioned home, a small system may be all that is needed. There is no sense in paying $10,000 for a back-up heating system that

---

**HEATING SYSTEM OPTIONS**

*Forced-air systems* distribute warm air to rooms via ducts with heat supplied by furnaces, heat pumps, wood furnaces, or solar thermal systems.

*Radiant floor* and *baseboard systems* distribute warm water to in-floor tubing or heat radiators located along walls with heat supplied by boilers, heat pumps, wood furnaces, or solar thermal systems.

*Wood stoves* and *masonry heaters* are devices that burn wood to produce heat.

*Wall-mounted heaters* burn natural gas or propane to produce heat.

---

provides $100 to $200 worth of heat per year. In addition, many people want systems that produce comfortable heat and operate quietly. Systems must also be reliable and should be easily serviceable.

Another consideration to keep in mind when selecting a supplemental heating source is your heat-delivery requirements. Will you need a sudden burst of heat for a day or two, every other week during a cold spell, or will you need a system that provides a steadier amount of heat, for example, one that operates most nights during the dead of winter? Will you need a lot of heat or just a little?

Automation is another key criterion. Do you want a system that will operate automatically when you are away to protect pipes from freezing? Some systems require a fair amount of operator involvement and are not therefore suitable for those who travel a great deal.

To complement the environmental benefits of a passive solar home, you will no doubt want to install an energy-efficient system, one that produces the maximum amount of energy from the fuel it burns, and produces as little pollution as possible. You may want to install a system powered by abundant, renewable fuels.

In this chapter, we'll explore the major options for back-up heating, beginning with forced-air heating systems, proceeding generally from the least sustainable to the most sustainable. I will briefly describe how they work and then detail the pluses and minuses of each option, including health and safety issues. I've also included a table that compares the various systems on key criteria to help make the task of selecting a back-up heating system for a passive solar home a little easier (table 4-1).

As you will soon see, judging the sustainability of back-up heating sources is rather difficult. As a rule, the more renewable and cleaner the fuel the more sustainable a system is. But of equal importance is the efficiency of the system and the amount of material that goes into components of a system. Even though a solar hot water system uses an abundant, clean fuel, it requires panels, pipes, tanks, and sometimes duct systems. A simple wood-burning stove might be a more sustainable option.

Whatever system you select, it is helpful to make your decision early on. Every type of heating system requires some up-front planning for optimal installation and performance. If you're going to install a solar hot-water system, for example, you'll need to accommodate the pipes that transport the heat transfer fluid from the rooftop to the water storage tank in your basement or utility room. It is much easier to plumb a system like this before your walls are finished. If you don't make up your mind early on, your options may be limited or costs may escalate wildly. Even if you cannot afford to install the system of your choice, plan ahead. Build the house so it can be added economically and with as little effort as possible at a later date.

## TABLE 4-1. Comparison of Back-Up Heating Systems

| TYPE | RESPONSE TIME | RENEWABLE FUEL | POLLUTION | CAPACITY | COMFORT LEVEL | EFFICIENCY | COST | RANK 1-10 (1 = poor ; 10 = great) |
|---|---|---|---|---|---|---|---|---|
| Fireplace | Fast | Yes | High | Room heating | Low | Extremely low | Moderate | 1 |
| Wood Stove | Fast | Yes | Low | Room heating or small houses | Low | Medium to high | Moderate | 7 |
| Pellet Stove | Fast | Yes | Low | Same as wood stoves | Low | Medium to high | Moderate | 7 |
| Masonry Heater | Slow | Yes | Low | Same as wood stoves | High | High | High | 9 |
| Forced Air (gas or oil) | Fast | No | Gas–low Oil–medium | Whole house | Medium to high | Medium to high | Moderate | 5 |
| Radiant Floor (gas) | Slow | No–unless supplied by heat pump or solar hot water | Low | Whole house but capable of zone heating | High | Medium to high | High | 9 |
| Base-board Hot Water | Medium | No–unless supplied by heat pump or solar hot water | Low | Same as radiant floor | High | Medium to high | High | 8 |
| Heat Pump | Fast | Partially. Heat is renewable but electricity to run them is generally not | Low | Same as radiant floor | High | High | High | 9 |
| Solar Hot Water | Medium | Yes, except for electricity to run pumps | Low | Same as radiant floor | High | High | High, including radiant-floor system to distribute heat | 9 |
| Electric Baseboard | Fast | No | High | Whole house and room heating | Medium– can produce dry indoor air | Low | Installation is low, but operating cost is high | 1 |
| Wall Heater (gas) | Fast | No | Medium | Room heating | Medium, tends to produce hot zones | Medium | Low | 7 |
| Wall Heater (electric) | Fast | No | High | Room heating | Medium | Low | Low | 1 |

## FORCED-AIR HEATING SYSTEMS

According to the U.S. Department of Energy, 62 percent of all homes in the United States are heated by forced-air systems. Heat in such systems can be generated by any of several different devices: gas- or oil-fired furnaces, electric furnaces, wood furnaces, electric furnaces, heat pumps, and solar heaters (solar hot-water systems).

Warm air generated by these sources is then circulated through extensive ductwork located in floors, walls, and ceiling cavities (figure 4-1). A large, fairly powerful fan in the furnace is required. Hot air flows through the ducts, then escapes via registers into the various rooms of the house. It circulates through the rooms, cooling as it goes, then returns to the furnace in a separate set of ducts, the cold-air return.

### Pros and Cons of Forced-Air Systems

Forced-air systems provide immediate heat, and are popular for this reason. In addition, these systems can be designed to serve additional functions. The ducts can, for instance, distribute cool air from a central air conditioner during hot, summer months. They can also be integrated into a ventilation system, discussed in chapter 6. Humidity and dehumidification features can be added to the furnace and ductwork. Air filters can be installed in the system to remove dust, pollen, and other indoor air pollutants. Finally, the circulating fan can be used to help distribute heat generated in a passive solar home. During the day, for instance, the air handler can be left running to move air throughout the house for a more even distribution of heat. Because forced-air systems are so common, finding a qualified person to install or repair one is not difficult

Perhaps the biggest drawback of forced-air systems is that they are generally the least efficient of any heating system. This is not so much because of the furnace, but rather because of the continual movement of air that can pressurize the inside of a home, forcing warm air to escape through numerous cracks and crevices in the building envelope. The more warm air that is forced out of a house, the higher the heat bill. Ducts can leak, too, spilling heat into unintended spaces. Air flowing out of vents in the system also produces uncomfortable drafts. As a result, the thermostat setting may need to be higher in a home with a forced-air system to obtain the desired level of comfort.

Forced-air systems are also not amenable to zone heating. Although you can shut registers off in unused rooms, chances are that much more of a house will end up being heated during the winter than might

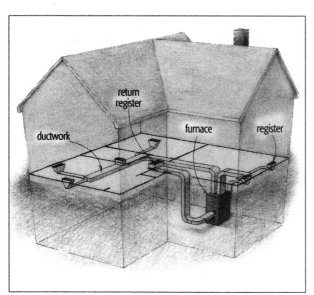

return register

ductwork

furnace

register

otherwise be heated with other systems, for example, a radiant-floor or base-board hot-water system.

Moving air can also stir up dust, including dander from people and pets. For people with allergies this can be a major problem. Duct systems can also serve as breeding grounds for mold. When the fan is running, spores can be dispersed throughout the house.

A final disadvantage is that many forced-air systems are noisy. The fans required to move air throughout an entire home are rather large, and generate a fair amount of racket.

**BUILDING NOTE**

When installing a forced-air system be sure to insulate ducts that run through unheated spaces, for example, attics or crawl spaces. Also be sure to seal all ducts to prevent heat spillage into unheated spaces.

## BASEBOARD HOT-WATER SYSTEMS

Baseboard hot-water systems produce a clean and relatively comfortable form of heat. Like forced-air systems, baseboard hot-water systems (also known as hydronic baseboard heating systems) require a central heat source, usually a boiler, to produce hot water. Hot water is pumped through a network of copper pipes in the walls and floors to small radiators located along the junction of the wall and floor (figure 4-2). Within the radiators aluminum fins disperse the heat. A covering is placed over the fin for aesthetics.

Baseboard heating transfers heat to rooms primarily by convection: air currents created by heat emanating from the baseboard units. As air surrounding the baseboard units is heated, it rises. Cool air gently moves in to replace it, so there are no uncomfortable drafts as in forced-air systems.

### Pros and Cons of Baseboard Hot-Water Systems

Baseboard hot-water systems produce extraordinary comfort because, in virtually all homes, the baseboard heaters are placed along the outside of the walls. This arrangement shields occupants from cold surfaces, such as windows and sliding glass doors. Occupants of a home are surrounded by a blanket of warmth.

Another advantage is that baseboard hot-water systems are amenable to zone heating. Each zone can have its own thermostat, so unused rooms can be turned off or turned down when not in use. Any or all of the zones can operate at one time.

Radiant baseboard systems are relatively easy to install in new construction and can also be installed in existing homes. In addition, baseboard hot-water systems are relatively quiet and reliable, provided you install a high-quality boiler. Even so, well-made systems do require occasional maintenance.

Hot-water baseboard is a fairly inconspicuous system. Baseboard units are slim and architecturally neutral. They are also available in several colors, even

FIGURE 4-2
*Hot water produced by the boiler circulates throughout the house and releases its heat at strategically placed baseboard heaters (radiators). Drawing by Christopher Clapp. Reprinted with permission from* Energy-Efficient Building, *© 1999 by the Taunton Press, Inc.*

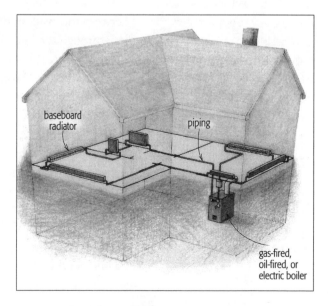

baseboard radiator

piping

gas-fired, oil-fired, or electric boiler

## INSULATE SLABS WHEN INSTALLING RADIANT-FLOOR HEAT

When installing a radiant-floor heating system, be sure to insulate underneath the floor and around the perimeter of the foundation. John Siegenthaler, who designs systems for northern New York State, argues that "Adequate insulation is an essential component of any slab-on-grade system." He, like others, uses high-density expanded polystyrene, which is not made using ozone-depleting hydrochlorofluorocarbons, under the slab and along the perimeter to thermally isolate the slab from the Earth (figure 4-3). Failure to do so wastes a large portion of the heat supplied to the slab. Overinsulating a slab is better than underinsulating; if you underinsulate your slab, you can't go back and fix it! Manufacturers typically recommend one-inch rigid foam insulation. Two to four inches is best.

wood grain, to complement a home's decor. Because the surface temperature of baseboard units is not very high, compared to electric baseboard, furniture and drapes can come in contact with them without fear of fire.

Baseboard hot-water systems are efficient and do not contribute to indoor air pollution. That is because most systems are installed with high-efficiency boilers with closed or sealed combustion chambers and an outside source of combustion air (described below). High-efficiency boilers reduce fuel bills and emit fewer air pollutants than less-efficient models.

The efficiency of a hot-water baseboard system is also enhanced because, unlike forced-air systems, air is not blown around creating a positive pressure inside the house and forcing warm air out through cracks in the building envelope. Baseboard hot-water is also a draft-free form of heat.

Although they are an excellent choice, baseboard hot-water systems have some disadvantages. One of the most significant is that these systems warm a room more slowly than forced-air systems.

In addition, boilers typically burn a nonrenewable fuel (usually natural gas or propane). The combustion of these fuels adds carbon dioxide and carbon monoxide to the atmosphere. Natural gas supplies are also limited. The United States has consumed over half of its domestic supplies, and natural gas is not a resource that is easily imported. The U.S. currently acquires about 1 percent of its natural gas from Canada through pipelines. Importing from overseas suppliers in a compressed liquid form is costly and supplies are not likely to increase substantially.

One of the biggest disadvantages of baseboard hot water is the way the systems are installed, along the perimeter of a room and often under windows. Although this produces maximum comfort, it wastes considerable amounts of fuel. Warming windows and exterior walls increases the temperature difference between the wall and the outside. The greater the temperature difference between the inside and outside surfaces of an exterior wall, the greater the heat loss through the building envelope.

Baseboard hot water is also rather pricey, especially in retrofits, unless pipes can be run through a basement or a crawl space. (Be sure to insulate a crawl space if you go this route.) Even then, running copper pipes to second-story rooms can be quite challenging and costly. For an energy-efficient passive solar home, the cost of a system that is required to provides so little heat may not be justifiable. Finally, baseboard hot-water systems are single-purpose systems, unlike forced hot-air heating. That is, they cannot be used to cool, ventilate, or filter air.

## RADIANT-FLOOR HEAT

Radiant-floor heating systems are often used as a source of back-up heat in passive solar homes. Two types are available: electric and hydronic.

In electric systems, the less common of the two, heat is produced by running electricity through wires installed in the slab. Although it is a quiet source of heat, because it requires no furnace, electric resistance heat is extremely expensive to operate. Moreover, electricity is primarily generated in coal-fired or nuclear power plants. Neither power source scores high environmentally.

Most radiant-floor systems are hydronic; that is, they rely on heated water that is pumped through pipes in the floors of a house. Most hydronic radiant-floor systems are installed in an insulated concrete slab (figure 4-3). However, new materials and installation techniques have increased a builder's options. As a result, radiant floor can also be installed on wood-frame floors, under either tile or wood (figures 4-4 and 4-5). Even so, radiant floor in concrete slabs is the most economical application of all because the installed concrete slab is already factored into construction costs. The only additional cost is for installing the tubing and subslab insulation.

Radiant-floor systems consist of five components: (1) a heat source, usually a high-efficiency boiler, which produces the hot water; (2) manifolds, which help distribute heat to various parts of the house; (3) one or more pumps; (4) tubing laid in the floor; and (5) controls.

Most homes with radiant-floor heat contain two or more circuits, which permit zone heating. This

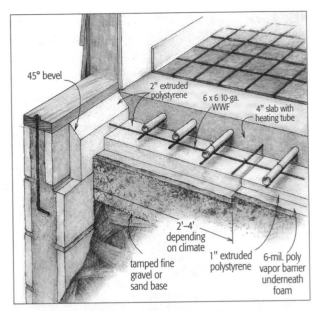

FIGURE 4-3. *In a radiant-floor system installed in a concrete slab, heat is distributed through pipes embedded in the concrete. Insulation below the slab and around its perimeter reduces heat loss to the ground, a significant problem in cold climates.*

FIGURE 4-4 (below left). *Radiant-floor heat can be installed over wood-frame floors by pouring a thin slab over the pipes, then laying tile over the slab.*

FIGURE 4-5 (below right). *Radiant-floor heat can be installed beneath wooden floors by using aluminum plates attached to the subflooring.*

*Drawings by Christopher Clapp. Reprinted with permission from* Energy-Efficient Building, © 1999 by the *Taunton Press, Inc.*

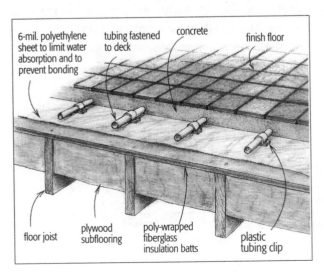

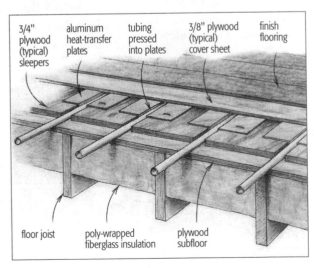

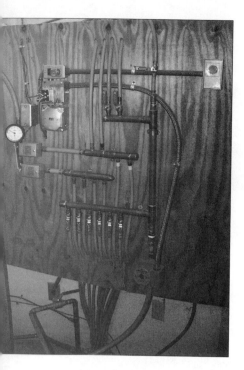

feature helps contribute to the efficiency of a radiant floor system: There is no sense in heating an entire home if you're spending most of your time in the living room and bedroom.

Although copper tubing was used in earlier systems, most installers now use cross-linked polyethylene (PEX) tubing. It is more flexible, durable, and easier to install than copper, and less prone to leakage.

Water heated by the boiler is delivered to the manifold, shown in figure 4-6. The manifold is a distribution center that delivers hot water to various circuits in the floor, where it gives off its heat. After flowing through the various circuits, the water returns to the manifold, from which it is pumped to the boiler for reheating.

## Pros and Cons of Radiant-Floor Heat

Radiant-floor heating was first used in ancient China and Rome about 60 AD in bath houses and homes (figure 4-7). Heat from fires was channeled through plenums in floors to a chimney in the walls. Warmth from the fire's smoky exhaust was absorbed by the floor and then radiated upward into the room.

Modern radiant-floor heating has come a long way since the days of the Romans. Today, these systems are a fairly efficient and relatively clean way of providing comfort, if sized and installed correctly.

Comfort is probably the most important benefit of a radiant-floor system. Unlike forced-air heating systems, which heat the air inside a house, radiant heat travels through the room air without warming it. It's heating effect occurs when infrared radiation, heat, strikes solid objects, for example, walls, ceilings, chairs, tables, and people. With such a large surfaces radiating heat to people and objects, radiant heating systems result in a higher mean radiant temperature—that is, a higher surface temperature in a space. The warmer the room surfaces, the less heat loss occurs from occupants. The result: People are more comfortable, even at lower thermostat settings.

A heated floor feels neutral to slightly warm on bare or stocking feet. Radiant-floor heat is very gentle, and avoids those annoying drafts you find in a home heated with a forced-air system. For those who love hard-surface

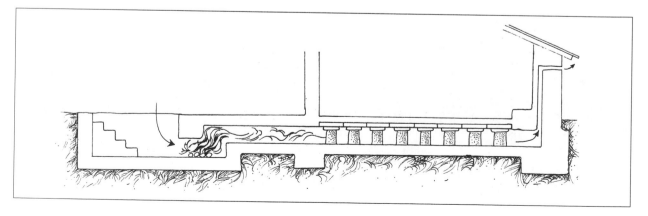

floors, such as tile, bamboo or wood, radiant-floor heat can be a great asset. "There's nothing worse than getting up in the morning and walking barefoot across the ice-cold kitchen floor to get to the coffeepot," notes Christine Grahl in an article on radiant-floor heating in *Environmental Design and Construction*. "No matter how warm the house is, the floor feels like a frozen pond."

Radiant-floor is the most efficient type of mechanical heating system. Overall, these systems consume 10 to 30 percent less fuel than their major competitors for the home-heating market: baseboard hot-water and forced-air systems. Why are they more efficient?

Unlike forced-air and baseboard hot-water systems, radiant-floor heating systems deliver heat to the floor, which then radiates into a room. As a result, radiant-floor systems deliver more of their heat to the room and its occupants than to outside walls, as with forced-air and baseboard hot-water systems, and hence to the chilly air outside a home. Efficiency is gained by pumps, too, which use less energy than the large fans required by forced-air systems. Lower thermostat settings and lower-temperature boiler settings also contribute to the efficiency of these systems.

Radiant-floor heat results in cleaner indoor air, too, especially if the boiler has a sealed combustion chamber. As Christine Grahl points out, with radiant-floor heat, "there is none of the blowing dust, drying air, and noise associated with forced-air furnaces."

As noted earlier, radiant-floor heat can be used in conjunction with many different types of floors. Hydronic heating systems tend to require less maintenance than forced-air systems, and they are easily serviceable. Unlike forced-air heating systems, there are no filters to replace.

Radiant heating is versatile in other ways, too. Like forced-air and baseboard hot-water heating systems, radiant-floor systems can be powered by natural gas, propane, oil, wood, electricity, and solar energy.

There are some drawbacks, however. Radiant-floor heating systems are expensive, costing $10,000 to $20,000, depending on the size of the house. As engineer and green building consultant Marc Rosenbaum points out, "It just doesn't make sense to put in a $10,000 heating system to provide $100 worth of heat per year."

Radiant-floor heat doesn't respond quickly. A room may take several hours to several days to reach a comfortable temperature. Systems installed in slabs are especially sluggish because they have so much mass to heat. If a homeowner turns the thermostat down for a couple of weeks while visiting relatives over Christmas, it may take a couple of days for the house to return to a comfortable temperature if the sun has not been available to heat the home. This problem can be avoided by keeping the thermostat set at a constant temperature; that way, the slab is always charged with heat during the heating season.

One of the biggest problems I have found is that radiant-floor heat transfers heat to objects in a direct path. When you are curled up on a couch reading

When it comes to selecting a boiler or furnace, many contractors err on the conservative side. That is to say, they tend to oversize them. Although this may seem like a logical choice, it is not. Oversized units are less efficient than properly sized units because they tend to cycle on and off more frequently than smaller, more appropriately sized furnaces and boilers. Start-ups use additional fuel and oversized heaters rarely operate long enough to reach optimal efficiency.

According to the U.S. Department of Energy, it is not uncommon for heating systems to be two or three times larger than they need to be. Resist the temptation to add extra Btus just in case. As a rule, a heating system should be no more than 25 percent larger than the calculated heating load.

When sizing a furnace or boiler or having someone else run the numbers, remember that a well-designed passive solar home needs less heat than a conventional home—sometimes much less. Don't let a heating contractor talk you into a large furnace or boiler without running the heating load estimates.

at night, and therefore blocked from radiant-floor heat, you may feel a bit cold. (I find this weakness to be significant. I'm warm as long as I'm up and about, but when I sit down, I feel cold.) Although this is a major drawback, there are ways around the problem. One of the most effective is to install tubing in walls, so it radiates onto people while sitting down or snoozing on the couch. Some manufacturers even sell radiant-wall panels for this application.

A couple of final considerations: Unlike forced-air systems, radiant-floor heat cannot be designed to provide ventilation or air filtration. And leaks in tubing, although rare, can be costly to repair, especially in slab-on-grade applications. Leaks can cause a lot of damage if not detected early.

## ALTERNATIVE HEAT SOURCES FOR MECHANICAL SYSTEMS

Forced-air, radiant-floor, and hot-water baseboard heating systems operate automatically using thermostats, and are therefore popular among architects, builders, homeowners, and building department officials. However, these systems typically burn fossil fuels, either natural gas, propane, or oil to generate heat. They also use a fair amount of electricity to power pumps and fans that transport heat throughout a home, and they require extensive networks of ducts and pipes as well as a furnace or boiler, all of which consume a lot of natural resources in their manufacture. For these reasons, mechanical systems rank low on the sustainability scale. However, there are ways to improve the environmental performance of mechanical heating systems, including high-efficiency furnaces and boilers, heat pumps, and solar hot-water systems.

### High-Efficiency Boilers and Furnaces

Since the 1970s, many manufacturers have begun to produce efficient, clean-burning boilers and furnaces. Today, so many are available that the choice can become quite perplexing. To help you make a selection, I will summarize some of most important features to look for. Before you begin reading this material, you may want to read the sidebar on sizing a back-up heater for a passive solar home.

#### Gas Furnaces

Many new homes come equipped with high-efficiency furnaces designed to minimize indoor air pollution. They are known as *induced-draft gas furnaces*, so named because they contain a fan that forces combustion gases out of the unit, usually through a vent pipe. The fan is required because these furnaces extract so much heat from the combustion chamber that the flue gases would not be buoyant enough to escape. Most of these models offer middle-level efficiency, ranging from 78 to 85 percent.

Even more efficient are the condensing-gas furnaces, so named because the heat exchangers in these units are so efficient that the flue gases cool down to the point that moisture (water vapor) in them condenses to form a liquid. Condensation of a gas gives off heat, and these models therefore achieve efficiencies ranging from 90 to 97 percent. Because so much heat is removed by the heat exchanger in these models, flue gases are vented through plastic pipe. Condensate is drained to a nearby floor drain.

Another feature to look for in a forced-air furnace is a sealed combustion chamber. This feature prevents dangerous combustion gases, such as carbon monoxide, from escaping into the house. Both condensing and noncondensing furnaces come with sealed combustion chambers.

### Gas Boilers

High-efficiency boilers are also available for radiant-floor and baseboard hot-water systems. Gas boilers come in many sizes and designs. Induced-draft models with sealed combustion chambers achieve efficiencies from 80 to 85 percent. However, very few boilers achieve efficiencies over 90 percent, in large part because manufacturers do not produce gas boilers with the condensing feature. The reason is that the temperature of the water returning from the zones for reheating is higher than the condensation point of the water in the exhaust gas. As a result, the exhaust gases cannot be cooled enough to condense water. No condensation translates into lower efficiency ratings and less heat output per Btu of gas burned.

### Oil Furnaces and Boilers

Fuel oil is also used to heat homes in some parts of the country. The oil is injected into a combustion chamber in boilers and furnaces through a nozzle, creating tiny droplets that mix with air to promote combustion. Many furnaces and boilers are designed to enhance turbulence, which boosts combustion efficiency.

Oil furnaces and boilers on the market today offer efficiencies ranging from about 78 to 82 percent. Although condensing models are available, they are not very common because fuel oil contains many more contaminants (such as sulfur) than natural gas or propane. Condensing out of the combustion gases, these contaminants produce a corrosive liquid that can damage the internal components of a furnace or boiler. Therefore, many contractors recommend against the slightly higher-efficiency condensing models.

Interestingly, very few oil furnaces and boilers come with a sealed combustion chamber. Experience has shown that cold outside air drawn into the combustion chamber of an oil-fired furnace or boiler reduces combustion efficiency and may impede start up.

### Wood Furnaces

Many people now heat their homes with wood burned in specially designed furnaces (figure 4-8). Wood furnaces are installed in basements, utility rooms,

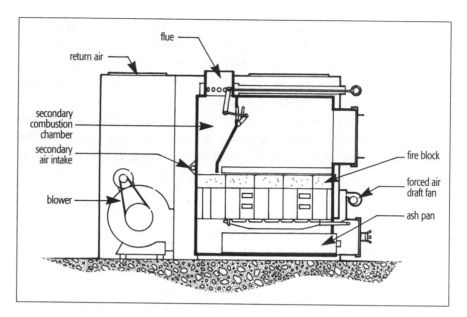

storage rooms, and even in garages, although there are some outdoor models. Systems are available for residential, commercial, and industrial applications with a wide range of heat output capacities.

Wood furnaces come in many sizes and many designs and are used in conjunction with radiant-floor and wall, baseboard, and forced-air heating systems. The Lynndale wood furnace, for instance, contains a blower fan that moves air through conventional ductwork (figure 4-8). The manufacturer, Lee Daniel of Harrison, Arkansas, also offers a hot-water option, designed to preheat domestic hot water. In Daniel's furnaces, the fire is controlled in part by a combustion air blower (a fan) located beneath the loading door, which blows a stream of air into the fire, improving the furnace's efficiency. This unit, like others, also features a secondary combustion chamber, which increases the efficiency by burning gases released from the wood during combustion. Increased efficiency translates into lower wood consumption and less pollution. In another model, the Lynndale furnace, the secondary combustion chamber has its own air supply. This feature promotes greater combustion efficiency and cuts down on air pollution and creosote building up in the chimney.

One important feature of wood furnaces is that combustion is electronically controlled. Here's how one manufacturer explains the system: When the thermostat calls for heat, the combustion air blower is activated and the fire is fanned until the heated air reaches a specified temperature. Then the air blower is shut off, conserving fuel and avoiding overheating often associated with less sophisticated wood furnaces.

Some wood furnaces are also designed to be added to an existing central heater. Summeraire, a company in Peterborough, Ontario, for instance, manufactures a wood furnace that can be used in conjunction with oil, gas, and propane furnaces. Another model burns wood and oil, creating a reliable supply of heat when you're away from your home for any length of time.

Wood furnaces have many of the same advantages and disadvantages of wood stoves. They are generally pretty efficient and clean burning. They use a renewable resource and they're fairly easy to operate. However, they do generally require large fans to distribute air through duct work. Fans can be noisy and, of course, consume electricity. Loading wood and removing the ashes can be a chore, too. They also present some fire hazard, if not properly installed and maintained.

Outdoor models are an interesting variation on the wood furnace. Central Boiler in Greenbush, Minnesota, manufactures a line of high-efficiency outdoor furnaces made from heavy-gauge stainless titanium-strengthened steel (figure 4-9). These shed-sized furnaces can be placed as far as 500 feet away from a home. They burn wood to heat water, which is then pumped to the home through buried insulated pipe.

Outdoor furnaces can serve as the heat source for a variety of indoor heat distribution systems, including radiant-floor, baseboard hot-water, and forced air. They can also be used to supply domestic hot water.

Another sturdy, well-made outdoor wood furnace is manufactured by Heatmor in Warroad, Minnesota. They come in a variety of attractive colors to match the house or to blend into the environment. These furnaces are easy to load and maintain, and efficient to operate. They use a heavy-gauge steel that the manufacturer claims long outlasts its competitors.

## Heat Pumps

A heat pump is a device that extracts heat from the ground or the air around a home and transfers it to the interior of a building. They can be used as the heat source for any of the conventional back-up heating systems so far discussed, and can be used for summer cooling as well. Heat pumps come in two varieties: air-source and ground-source.

*Ground-Source Heat Pumps.* Ground-source heat pumps extract heat from the earth around a home. As illustrated in figure 4-10, a heating system that relies on heat pump consists of three parts: (1) pipes buried in the ground to draw heat from the earth, (2) the heat pump itself, and (3) some means of distributing heat in a house (radiant floor or forced air, for instance).

In the winter, ground-source heat pumps gather heat from the subsoil, beneath the frost line, through an extensive set of buried tubes. Water pumped through the buried tubing gathers heat from the ground. The heat is then concentrated by the heat pump and transferred into a home.

FIGURE 4-9
*The Central Boiler outdoor wood furnace reduces the need to haul wood into a house and removes the fire danger as well.*

*Ground-source heat pumps do not generate heat like furnaces and boilers, they simply transfer it from one place to another. Because they do not burn a fuel, these devices produce no indoor air pollution.*

FIGURE 4-10

*Components of a heat-pump system. Pipes are laid horizontally (a) in the ground four to six feet below the surface or vertically (b) where they extend 100 to 400 feet below the surface. Water or a mixture of water and an environmentally benign antifreeze circulates through the pipe, gathering heat in the winter. A small electric pump provides the propulsive force.*

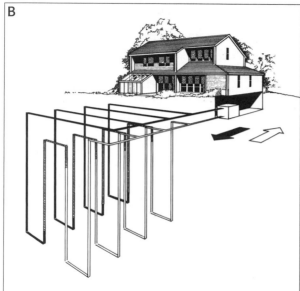

**DOMESTIC HOT WATER BONUS**

Many residential ground-source heat pump systems come with a device known as a *desuperheater.* It transfers waste heat from the pump's compressor to a hot-water tank when the unit is running, thus providing free hot water. In the summer, the units provide 100 percent of a family's hot water. In the winter, they typically satisfy about half the hot water demand.

Heating a home in this manner is possible for a number of reasons, but primarily because the subsoil remains at a fairly constant temperature year-round—around 50°F, give or take a little depending on where you live.

Ground-source heat pumps rely on refrigeration technology—refrigerant gases, compressors, and pumps—to extract heat from the ground. The heat delivered to the unit from the underground pipes causes the refrigerant in the heat pump to expand. Expansion of the refrigerant releases heat, which is then captured and transferred to the heating system. The refrigerant gas is recycled—it is recompressed—so it is ready to repeat the cycle.

Ground-source heat pumps can also be used to cool homes by running in reverse during the summer. In summer, the device gathers heat from the interior of a house and releases it into the ground. A few valves in the system allow heat-exchange fluid in the system to follow these two different paths, one for heating and one for cooling, thus eliminating the need for separate furnace and air-conditioning systems.

Heat pumps can also be used to draw heat from groundwater or surface water, for example, ponds or lakes. The water is then pumped back to the source or disposed of on the surface. These units are known as water-source heat pumps.

*Pros and Cons of Ground-Source Heat Pumps.* Ground-source heat pumps are an energy-efficient means of heating and cooling homes. In fact, the U.S. Department of Energy and the Environmental Protection Agency (EPA) both consider the ground-source heat pump to be the most efficient, environmentally benign, and cost-effective space heating and cooling system on the market today. These systems use relatively small amounts of electricity to power pumps and compressors—about 25 to 50 percent less electricity than conventional heating and cooling systems. If electricity is supplied from a renewable source, the environmental gains are even greater. Moreover, according to the EPA, ground-source heat pumps offer the lowest carbon dioxide emissions of any mechanical heating and cooling systems.

Ground-source heat pumps can be used in virtually any climate. Although they're more expensive to install than many other heating and cooling systems, efficiency gains pay for the additional costs in two to ten years. Moreover, units carrying EPA's Energy Star label can be financed with special Energy Star loans from banks and other financial institutions. Some of these loans carry a lower interest rate, others allow larger loan balances, and some combine both features. (For information on Energy Star loans, call 1-888-STAR-YES.)

Ground-source heat pumps can be installed in new homes and as a retrofit and are more compact than conventional heating and air conditioning systems. Because they have relatively few moving parts, they require less maintenance than conventional heating and cooling systems. Underground piping is often warranteed for twenty-five to fifty years.

Perhaps the greatest advantage of the ground-source heat pump is that it relies on a renewable resource, solar energy. Sunlight heats the upper surface of the ground and is the energy source for the system. Because these units contain no source of combustion, they produce no climate-altering carbon dioxide or other air pollution other than that involved in the generation of the electricity consumed. They produce no indoor air pollution. No combustion also eliminates the risk of house fires. Ground-source heat pumps also operate fairly quietly.

Like every other system for providing heat, ground-source heat pumps have their disadvantages. The main problem is that these devices use a refrigerant known as *hydrochlorofluorocarbon–22* or *HCFC-22*. Although this chemical is less stable than ozone-depleting CFCs and therefore tends to break up in the lower atmosphere, it does reach the ozone layer and destroys ozone molecules. As you probably know, ozone provides a protective shield against ultraviolet B radiation, which causes cataracts and cancer, and injures plants. Although HCFCs destroy far fewer ozone molecules than CFCs, 5,000 compared to 100,000, ozone lost is still significant. Fortunately, ground-source heat pumps come with factory-sealed refrigeration systems that will seldom or never have to be recharged, reducing potential leaks and ozone destruction.

Installing horizontal and vertical loops in a ground-source heat pump system can cause significant damage to a building site, and will require special efforts to prevent erosion and to ensure revegetation.

While efficient, ground-source heat pumps do cost substantially more than conventional heating and cooling systems, largely because they require extensive excavation. According to the Geothermal Heat Pump Consortium, an industry group, these systems cost $2,000 to $5,000 more than a conventional

*According to the EPA, ground-source heat pumps use 44 percent less energy than air-source heat pumps and 72 percent less energy than an electrical resistance heater with a standard air conditioning unit.*

## IMMEDIATE MONETARY SAVINGS

Energy Star loans for ground-source heat pumps (and other energy-efficient heating systems) can result in immediate positive cash flow. For example, if a heat pump adds $25 to the cost of a mortgage, but saves $30 per month in utility bills, the homeowner is ahead of the game. Additional savings can be achieved because mortgage payments can be deducted from income taxes, whereas home energy bills cannot.

## HOW COST-EFFECTIVE IS A GROUND-SOURCE HEAT PUMP?

As a general rule, a ground-source heat pump only makes sense for heating-load dominated buildings— that is, buildings whose annual heating bill is equal to or greater than its cooling bill—when the cost per kilowatt-hour of electricity is less than one-tenth the cost per therm of natural gas. For example, if natural gas costs 80 cents per therm, electricity would have to cost less than 8 cents per kilowatt-hour for a system to be economical.

NADAV MALIN AND ALEX WILSON, "Ground Source Heat Pumps: Are They Green?" in *Environmental Building News.*

heating and cooling system. In a well-designed passive solar home, the additional cost for back-up heat may not be worth it (see sidebar). Contact your local utility for rebates or other incentives when considering this type of system.

Finally, on the issue of impacts, open-loop water systems that draw groundwater to the heat pump can deplete aquifers. Dumping groundwater on the surface may cause problems, too. If you want to read more about them, see Nadav Malin's and Alex Wilson's article "Ground Source Heat Pumps: Are They Green?" in *Environmental Building News.*

*Air-Source Heat Pumps.* Air-source heat pumps are another option. They operate the same way as a ground-source heat pump, but capture heat from the air rather than the ground. In the winter, an air-source heat pump captures heat from the outside air and moves it indoors. In hot weather the operation

FIGURE 4-11

*An air-source heat pump consists of indoor and outdoor portions. After giving off its heat inside a house, the cool refrigerant passes from the house to the outside. Here the pressurized liquid enters an expansion device and is converted to a low-pressure, low-temperature liquid, which then enters the outside coil. A fan blows outside air over the cold coils. Heat is transferred from the outside air into the coils where it warms the refrigerant, transforming it from liquid to gas. As the gas expands, it absorbs heat. Next, the heated refrigerant gas passes through a compressor, reducing its volume. The heated, high-pressure refrigerant then enters the house. Because it is hotter than the inside air, it gives off heat in the inside coil. A fan blows across the coils, stripping the heat away. When cooling a house in the summer, the process is identical, except the heat is obtained from inside the house and transferred to the outside.*

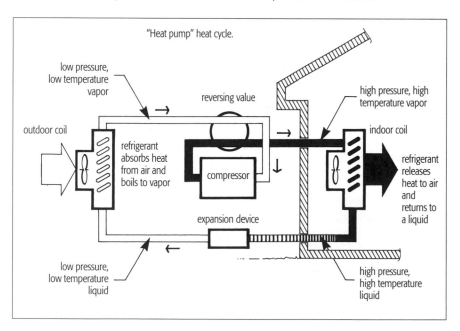

switches, stripping heat from the house and dumping it outside. These are currently by far the most widely sold model.

As illustrated in figure 4-11, the air-source heat pump consists of an indoor and an outdoor portion. The outdoor portion consists of coils filled with a refrigerant that absorbs heat from the air, even at low temperatures. Because the refrigerant is cooler than ambient air, sometimes as cold as 0°F, it can absorb heat from it.

Manufacturers sell many more air-source heat pumps than ground-source heat pumps. However, don't construe that to mean that they are better. Air-source heat pumps are not as efficient as ground-source heat pumps. Nor are they as widely applicable. As a rule, they work best in warm climates, with significant cooling loads and minimal heating demands. Small amounts of heat can be provided when temperatures drop, but if the air temperature falls consistently below 35°F, an air-source heat pump will require a supplemental heat source. In addition, air-source heat pumps produce more carbon dioxide (due to the combustion of coal to make electricity) than an efficient furnace or boiler burning natural gas.

Of the two models, choose a ground-source heat pump. They use half as much HCFC and deliver more heat (or cooling) per unit of electricity consumed than air-source heat pumps.

## Solar Hot-Water

The sun that heats a house passively can also be used to supply hot water for back-up heat. Solar hot-water systems for space heating consist of solar collectors, which are typically mounted on the roof of a house, a system of pipes, pumps, and a storage tank and heat exchanger (figure 4-12).

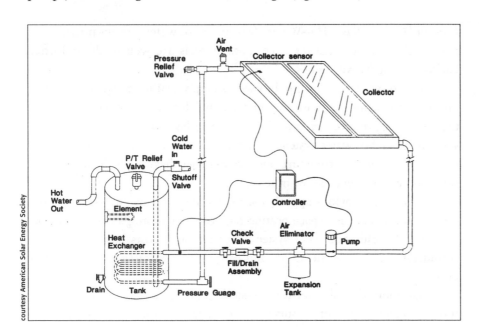

courtesy American Solar Energy Society

**FIGURE 4-12**
*One of the most common active systems is known as a closed-loop system. This system contains a heat-exchange liquid, usually a mixture of water and a relatively safe form of glycol antifreeze, which circulates through the panels and back to the tank, where it gives off the heat it has acquired in a heat exchanger, usually a series of pipes in the wall or in the base of the water tank. After giving off its heat, the fluid is pumped back up to the roof to be reheated.*

For years, the most commonly used solar panels consisted of flat rectangular boxes, known as *flat-plate collectors*. With an interior black surface, a pane of glass over the front surface to let light in, and well-insulated sides and bottom, the solar panel gathers up sunlight and converts it to heat. Interior temperatures can easily climb to well over 200°F. Heat created inside the panel is drawn off by a fluid transported through pipes located in the interior of the box. These pipes lead to a water storage tank, typically located inside the house, usually in a basement or a utility room. Pipes connecting the collectors and storage tank serve as conduits for the circulation of a heat-exchange fluid from the water storage tank to the collectors and back again. The heat exchanger allows heat to be transferred into the water storage tank. Heated water is then used to provide heat for forced-air or radiant-floor heating system.

Solar hot-water systems designed for domestic hot water generally require tanks with capacities of 80 to 120 gallons. A system designed for heating a home could have a tank that holds up to 1,500 gallons. In either case, the tanks are usually extremely well insulated to retain heat. Because they have such a huge reserve capacity, they can serve as back-up heat for a superinsulated, energy-efficient passive solar home.

*Types of Solar Hot-Water Systems.* Solar hot-water systems operate quite simply, although some can be sophisticated. Two types are in use today, active and passive, with many variations on these basic themes.

An active system requires pumps and various control devices to propel the liquid heat-exchange medium through the system (figure 4-12). Passive systems operate without pumps. They rely on water pressure, gravity, or buoyancy (natural convection) to act as a pump.

*Pros and Cons of Solar Hot-Water Heat.* Solar hot-water systems provide heat from a clean, renewable resource. New models are well made and often backed by lengthy warranties.

Solar hot-water systems can be used to supply heat for conventional back-up heating systems—that is, radiant-floor and forced-air systems (although they may not produce hot enough water for a baseboard hot-water heating system). Solar hot-water systems can also serve to heat water for domestic uses. If you've got a swimming pool, excess solar heat could be used in the spring or fall to extend the swim season.

Despite their obvious appeal, solar hot-water systems have some serious drawbacks. Systems for space heating require many more panels than a conventional domestic hot-water system. And although costs have come down considerably since the late 1970s and early 1980s, these systems tend to be fairly pricey.

Installation of a solar hot-water system can be difficult in a retrofit, where roof orientation may not be suited to solar applications. Active systems may

FIGURE 4-13
*The Thermomax solar collector gathers heat even on cloudy days.*

require periodic repair; sensors and controllers are the most common sources of breakdown. Moving parts in motors also require periodic maintenance, and maintenance means fixing things yourself or hiring someone to perform the work for you. Unfortunately, a single repair bill can erode the economic benefit.

For a conventional home, solar thermal systems for back-up heat don't make a great deal of sense economically, unless you are currently heating water for your system with electricity—an extremely expensive option in most places. For an off-the-grid home, solar thermal systems still tend to be much more expensive than a conventional water heater fueled by natural gas or propane, although you can get some great buys on used systems. Contact Real Goods or call a local solar dealer for assistance. You may also be able to purchase a system through the want ads. There's always someone trying to sell an old system. However, be careful what you buy. The generous solar tax credits offered in the late 1970s and early 1980s by states and the federal government resulted in "an abundance of poorly designed and sloppily installed solar hotwater systems, giving a lingering black eye to the whole solar industry for many folks," according to John Schaeffer, president and founder of Real Goods, a company that sells environmentally friendly products, including solar systems. Because you don't want to end up with any mistakes from the past, you will very likely be better off buying a new system, which will tend to be simpler and better made and therefore less likely to break down.

One of the most impressive new systems I've seen is the Thermomax solar water heater. This system consists of a series of parallel glass vacuum tubes, as shown in figure 4-13. Each vacuum tube contains a black copper pipe containing alcohol. When sun strikes the pipe, the alcohol is heated and hot vapor rises to the top of the tube by convection, where heat is removed by a heat exchanger located along the top of the unit. Heat is then pumped into the house and through another heat exchanger in the wall of the storage tank.

Thermomax is extraordinarily efficient, even in cloudy and cold weather. According to the company, it converts twice as much solar energy to heat as a conventional flat-plate collector. This system performs so well because vacuums are tremendous insulators. The absence of air molecules surrounding the alcohol-filled pipe greatly reduces the loss of heat. In a laboratory test on another similar model, a single 50°F water-filled tube placed in a –13°F freezer took 6.5 days to freeze. In Thermomax, ambient temperature can fall well below zero while the vapor inside the unit reaches temperatures as high as 300°F.

Thermomax also works well because the round design of the tubes is ideal for capturing diffuse radiant energy, as much as 80 percent of what's available during cloudy weather. A large system can provide ample back-up heat. The manufacturer claims that a family can meet 70 percent of its domestic hot water needs in worst-case climates. In areas with abundant sunshine you can meet 100 percent of your hot water demand with this system.

Thermomax is not the only efficient, reliable solar water heater on the market. Heliodyne PV-powered systems, sold by Real Goods and others, although more costly than many other units, offer the most trouble-free service in the long run, according to the technical staff at Real Goods. This system is a favorite of many solar installers, and the closed-looped plumbing can't freeze, even in the coldest climates.

## WALL-MOUNTED SPACE HEATERS

In superefficient passive solar homes that require very little or only occasional back-up heat, one of the most cost-effective systems is the wall-mounted space heater (figure 4-14). Numerous models are available to fit a wide range of applications. If necessary, two or three units can be installed in a home to provide heat where it is needed when it is needed.

Wall-mounted space heaters are great for heating infrequently used rooms, for example, a workshop or sewing room. They are also a valuable and economical back-up heat source for solar additions. Extending a home's heating system to supply an addition requires duct work (in the case of a forced-air system) or pipe (in the case of radiant systems). It may even require installation of a higher-capacity boiler or furnace or a second unit to service the add-on. All of these changes can be costly. A wall-mounted heater requires installation of only a gas line and possibly a vent pipe to remove combustion gases, although there are ventless models that eliminate this requirement.

Wall-mounted space heaters run on a variety of fuels, including electricity, propane, natural gas, and kerosene. Generally, the most efficient and best from an environmental standpoint are the propane and

FIGURE 4-14
*Wall-mounted space heaters are valuable for heating individual rooms or small houses, and are especially useful in solar retrofits.*

courtesy Rinnai

natural gas models, although these fuels are nonrenewable and their combustion produces pollutants such as carbon dioxide.

Wall-mounted space heaters operate by natural convection or by radiating heat into a room. Some units contain blower fans. Many come with built-in or external thermostatic control features, so the heaters automatically adjust to room temperature. This feature makes the wall-mounted heater suitable for heating a passive solar home when a family is away.

In a standard model, gases are vented via a flue pipe that extends either through the ceiling or directly through the wall. Some wall-mounted heaters come with an outside air intake feature, so that combustion air can be drawn from the outdoors rather than from interior air.

*Pros and Cons of Wall-Mounted Heaters.* Wall-mounted heaters have many advantages. They are widely available, inexpensive, reliable, easy to maintain, relatively inexpensive, and easy to install. Some are extremely efficient. Rinnai's vent-free models boast efficiencies of 99.9 percent. Wall-mounted heaters provide immediate heat and allow intermittent and automatic heating of selected spaces and, as noted earlier, are ideal for superefficient passive solar homes that require very little back-up heat or for solar additions. Two strategically placed wall-mounted heaters, costing $1,000 to $2,000 each installed, is a far better investment than the $10,000 to $20,000 required to install a radiant-floor heating system.

Wall-mounted heaters do have some drawbacks, however. For one, they are generally designed to heat individual rooms or small homes. Unlike radiant-floor or baseboard hot-water systems, they do not provide uniform heat. Like a wood stove, they tend to produce heat gradients with temperatures near the unit higher than temperatures on the far side of the room.

And then there's the issue of safety. The U.S. Consumer Product Safety Commission estimates that space heaters, including wood stoves, cause 25,000 house fires in the United States each year. More than 300 people die in these fires. In addition, 6,000 individuals are rushed to the emergency room each year with burns caused by touching space heaters. In addition to burns and fires, fossil fuel-powered space heaters may cause explosions and carbon monoxide poisoning.

Because of these problems, the Consumer Product Safety Commission recommends built-in safety features such as (1) an oxygen-depletion sensor that shuts off the heater when it detects reduced levels of oxygen, (2) a pilot safety valve that stops the flow of natural gas to a heater if the pilot light goes out, and (3) a thermal shut-off device that turns the unit off if it is not venting properly. Care must be taken when installing and operating a unit. A heater guard helps keep children and pets from contacting hot surfaces. Further recommendations for safe operation can be obtained from the U.S. Consumer Product Safety Commission, listed in the Resource Guide at the end of the book.

*Wall-mounted heaters can be designed to burn so cleanly that they do not require venting to exhaust pollutants, although skeptical building departments may prohibit them.*

Jeff Anderson

FIGURE 4-15

*A client of mine converted his leaky fireplace (left) into a cubby hole for his TV (right), a great use of an inefficient fireplace.*

## A GREAT DEAL ON A USED WOOD STOVE?

In older model wood stoves, most of the hydrocarbons driven out of the wood "go up in smoke" because the fire isn't hot enough to ignite them. These stoves operate in the 400° to 900°F range, well below the 1,100° to 1,300°F range in which hydrocarbons burn. What happens to the unburned hydrocarbons? They escape from the combustion chamber. Some of these gases condense on the cooler walls of chimneys and flue pipes, forming creosote. The rest escape into the atmosphere as unburned hydrocarbons. There, they react with sunlight and other pollutants to form photochemical smog.

## FIREPLACES, WOOD STOVES, AND PELLET STOVES

Many passive solar homes, including mine, rely on back-up heat from wood-burning devices: wood stoves, fireplaces, wood furnaces, and pellet stoves. These "systems" provide warmth and comfort, and depend on a renewable source of fuel: wood. Let's look at each of these options and key considerations.

### Fireplaces . . . Forget It!

Standard wood-burning fireplaces are aesthetic and appealing, and the sound of a crackling fire is always enticing. Unfortunately, fireplaces are extremely wasteful, achieving efficiencies of only 10 to 20 percent at best. In fact, most of the heat they produce is lost up the chimney. Thus, standard fireplaces are not recommended. The best standard fireplace I have ever seen was in a client's home. He closed it in and mounted his television in the beautifully detailed structure (figure 4-15).

If you live in a home with a fireplace, don't burn wood in it to stay warm. You are very likely wasting enormous amounts of energy. To improve the efficiency of an existing fireplace, however, you can install a fireplace insert. Inserts are specially designed wood stoves that come in a variety of sizes to fit into fireplace openings. Fireplace inserts use the chimney to exhaust gases and are often equipped with fans to transfer additional heat to the room. Many models have glass doors to retain the view of the fire. Shutting the doors reduces cold air flow into a house from the chimney when the unit is not operating. Finally, inserts are a lot more efficient than an open fireplace, boosting the combustion efficiency to about 40 to 50 percent.

### Wood Stoves

Wood stoves are a much better option than fireplaces. They are more efficient, produce much less pollution, and come in many shapes and sizes. In addition, most wood stoves are built to be fairly airtight, a feature that prevents combustion gases from escaping into the room.

Wood stoves come in a variety of styles but are typically constructed of a single layer of fire-resistant material, such as steel or cast iron. They consist of a main combustion chamber typically lined by fire brick, to prolong the life of the stove. Modern wood stoves permit the user to control the flow of air into the combustion chamber to regulate the combustion temperature and the length of the burn. The more air that is allowed into the combustion chamber, the hotter the fire and the shorter the burn time. Exhaust gases escape through a vent pipe. Access to the stove is through a sealed door, often with glass to allow a view of the fire.

To improve efficiency and reduce emissions from wood stoves, manufacturers include catalytic burners or secondary combustion chambers (figures 4-16 and 4-17). Like the catalytic converter in a car, the catalytic combuster completes the combustion of unburned hydrocarbons, wrestling more heat from the wood. It can improve wood stove efficiency by 10 to 25 percent.

Because they facilitate the combustion of gases and liquids given off by burning wood, catalysts can reduce creosote deposition on flue pipes by 80 percent or more. Creosote is a black, tar-like hydrocarbon that collects on flue pipes. It can catch on fire, and, when it does, creosote burns so intensely that the fire can spread to the home.

Although catalytic burners add to the price of a stove, they increase efficiency of combustion and reduce wood consumption, typically paying for themselves in about two years. The life expectancy of new catalysts is about three to six years, although experience in my area indicates they give out in a couple years at high altitudes.

Catalytic burners even work when the air supply to a fire is restricted to extend the burn time. In such cases, unburned hydrocarbons escaping from the combustion chamber pass through the superhot catalyst, burning nearly completely.

## HOW CATALYSTS WORK

A catalyst speeds up chemical reactions. It does so by lowering the temperature at which they take place. The body is full of catalysts called enzymes. In wood stoves, catalysts lower the ignition temperature of the unburned hydrocarbons to about 600°F. Normally, hydrocarbons given off by wood wouldn't burn until temperatures reached 1100 to 1300°F. Heat inside a conventional wood stove is usually 400 to 900°F. Catalysts initially burn hydrocarbons at a low temperature. But soon after they begin to work, the temperature inside the catalyst can climb to 1,700°F or more. Once a catalyst is up and running, it maintains temperatures sufficient to burn off hydrocarbons.

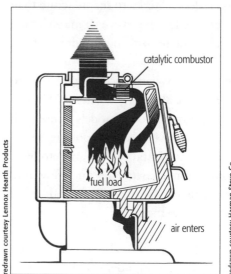

catalytic combustor

fuel load

air enters

redrawn courtesy Lennox Hearth Products

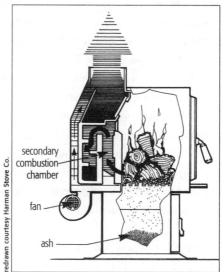

secondary combustion chamber

fan

ash

redrawn courtesy Harman Stove Co.

**FIGURE 4-16 (left)**
*Catalytic burners in wood stoves help reduce emissions of volatile gases and improve the stove's efficiency.*

**FIGURE 4-17 (right)**
*This wood stove contains a secondary combustion chamber that extracts additional heat by burning hydrocarbons released from the wood.*

**Catalytic**

- Generally highest efficiency
- Lowest emissions
- Longest burn times
- Higher cost
- Better suited to constant use

**Noncatalytic**

- Better flame for viewing
- No need to replace catalysts
- Lower cost
- Better suited to less-frequent use

## BAFFLED OVER BAFFLES?

Baffles in a wood stove increase combustion efficiency and reduce pollution by increasing air turbulence, which increases combustion efficiency, and by holding hot gases in the fire box longer. They also direct gases containing unburned hydrocarbons back into the fire so they will burn more fully and make the fire burn hotter. An insulated baffle keeps the temperature of the gases high to promote secondary burning.

British Columbia Ministry of Environment, Land, and Parks, "Reducing Wood Stove Smoke: A Burning Issue."

## WOOD = SOLIDS AND LIQUIDS

"Half to two-thirds of the fuel value of wood is locked up in gases and volatile liquids," according to David Lyle. They are hydrocarbons. When a piece of wood is burned, combustion of solid material (mostly cellulose) yields only 30 to 50 percent of the heat. The rest of the heat comes from the combustion of hydrocarbon liquids and gases given off when a log is heated.

Numerous manufacturers have designed their wood stoves to burn efficiently without catalytic burners. Some have inserted insulated baffles in the top of the combustion chamber. The baffles force combustion gases back over the flames, causing them to ignite and burn. These stoves achieve high efficiencies and low emission rates, almost equal to stoves equipped with catalytic combusters (see sidebar, page 151).

Prewarmed room air introduced into the combustion zone further enhances the efficiency of a wood stove. This is achieved by pipes or channels that run alongside the firebox, warming air before it reaches the fire. Preheated air helps to maintain a hotter fire, which promotes more efficient burning. Cold outside air can cool a fire and reduce its efficiency.

Most wood stoves fit into one of two categories: radiant or circulating.

*Radiant Wood Stoves.* Radiant stoves are made from cast iron, welded steel, or soapstone. These stoves warm space primarily through heat energy that radiates directly off the hot surface of the stove. However, heat is also acquired by air that circulates naturally over the surface of the stove and stove pipe, creating convection currents in a room that help distribute the warmth.

Most stoves sold today are radiant heaters. Their popularity stems primarily from their cost. Because they require less material to manufacture than the circulating wood stove, radiant heaters are less expensive to manufacture (although some cast iron radiant stoves can be quite costly due to more extravagant design—see figure 4-18).

Radiant wood stoves are not only relatively inexpensive (compared to other stoves and other heating options), they are also fairly efficient. Combustion efficiencies of 60 to 80 percent are common. Some models come with automatic controls that regulate air flow to achieve constant room temperatures. Vermont Casting's Encore wood stove, for instance, has an automatic thermostat that provides for steady room heating. It regulates air inflow into the stove and thus heat output. The stove is also designed so that wood can be loaded from the top without releasing smoke into the room.

*Circulating Wood Stoves.* Circulating wood stoves look like ordinary wood stoves (figure 4-19), but closer examination reveals a fundamental difference:

They are double-walled. In a circulating stove, the inner combustion chamber is constructed of cast iron or welded steel lined with fire brick. The outer shell is steel. Separating the two is a small air space.

When the fire burns inside the inner chamber, it gives off considerable heat to the inner shell. Room air circulating either passively or actively (thanks to a fan) through the air space between the shells draws heat away from the fire. The hot air is then vented into the room.

Although heat also radiates off the outer shell of a circulating stove, it never reaches the temperatures of single-walled stoves. For this reason, circulating stoves are a good choice for families with young children. Circulating stoves achieve efficiencies ranging from 60 to 80 percent.

*Pros and Cons of Wood Stoves.* Wood stoves are an excellent choice for back-up heat in a passive solar home, provided they are installed properly. They provide immediate heat. Moreover, wood stoves are widely available, relatively easy to use, and easy to install, although professional installation by a competent crew is highly recommended. Wood stoves are visually appealing, as well. Many models come with glass doors that allow a view of the fire.

Many people choose wood stoves for back-up heat because they are one of the least expensive systems on the market. Wood stoves do not require costly ducts or pipes to distribute heat, as in forced hot-air or radiant heating systems.

Wood stoves can be used to heat a room or an entire house, but they usually work better in smaller homes. Further adding to their appeal, wood stoves require little maintenance and cleaning (usually once a year), if properly burned.

FIGURE 4-18 (left)
*This cast iron stove by Vermont-based Hearthstone is not only efficient and easy to use, it looks great. Courtesy of Hearthstone Stores.*

FIGURE 4-19 (right)
*The circulating wood stove has a double-walled box that promotes circulation of air over the stove, stripping heat away from the inner layer. Courtesy of Hearthstone Stores.*

## THE ADVANTAGES OF SOAP STONE

Soapstone is used to build wood stoves because it provides additional mass. This helps retain heat, which is radiated into a room after the fire has gone out. Soapstone looks good, and while it takes a long time to warm up, it takes just as long to cool off.

## HOW BIG SHOULD YOUR STOVE BE?

To assist customers in their selection, some wood stove manufacturers provide data on heat output in Btus per hour. Others list the number of rooms or the room size (in square feet) each of their models is designed to heat. Still others list the cubic feet of room space a stove will heat.

Although this information is useful, unless you know the precise conditions (for example, temperature, insulation levels, and type of wood burned) under which the stoves were tested, these data may not help you much. Unless test conditions are identical to your home, the ratings are only approximations useful for comparing models manufactured by one company.

Your best bet is to ask the local supplier who is familiar with your area and heating requirements to give you a recommendation *after* carefully looking over your plans or visiting your home. Be aware, however, that many suppliers won't consider the added insulation and other details that make a passive solar home more efficient and more self-reliant than standard-built homes. You may end up buying a model that's much larger than needed.

FIGURE 4-20
*Make-up air replaces air drawn out of a room by a wood stove. Passive systems (no fans) are generally ill advised, as they afford no control over the process. Active systems allow much better control of air flow.*

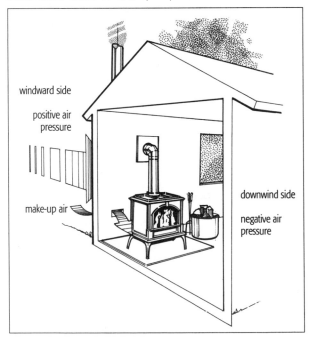

Wood stoves manufactured in the United States, Canada, and other countries where ambient air quality is a concern burn fairly cleanly. In fact, all wood-burning stoves sold in the United States and Canada, except cook stoves, must comply with government regulations. EPA-approved stoves burn so cleanly that they are exempt from wood-burning bans imposed in many cities and towns during high pollution days. Unlike their predecessors, which often produced black clouds of particulate-laden smoke, the new EPA Phase II stoves produce almost no smoke. Certified stoves reduce smoke emissions by as much as 90 percent compared to older models.

The state of Washington requires that all new wood stoves draw their combustion air from outside the home. This is called *make-up air*. Make-up air replaces air drawn out of a house while a fire is burning in a wood stove. It also ensures that a wood stove burns properly, even when a large fan such as a kitchen range exhaust fan is running.

Make-up air may be provided by creating a four-inch hole in the wall of a house, attaching a duct to the opening, then running the duct from the wall to the room containing the wood stove or to an adjacent room (figure 4-20). In some instances, the duct may be connected directly to the firebox so outdoor air flows directly into the combustion chamber. A low-wattage fan can be used to draw air, permitting the greatest control of air inflow with the fewest problems. For best results, make-up air should be pre-warmed, so as not to lower fire temperature.

To be effective and safe, make-up air systems must be carefully designed. Powered make-up air systems—that is, systems that draw air into a house using a fan—function best. For a detailed explanation of the challenges of make-up air, log on to the Wood Heat Organization's Web site (www.woodheat.org). I have also listed free software and on my own Web site that will allow you to simulate how a wood stove will operate in your house and the effects of indoor and outdoor air pressure on a stove's performance.

From an environmental perspective, wood stoves add no additional carbon dioxide to the atmosphere so long as new trees are planted or allowed to grow to offset carbon dioxide produced by burning the wood. Bill Eckert of Friendly Fire, a local wood stove distributor and past president of the Hearth Products Association, recommends that you plant a seedling of hardwood for every cord of wood you consume!

Also on the plus side, wood stoves rely on a renewable resource, which is abundant in many parts of the world. Sustainably managed wood lots can provide a lifetime of fuel to a family. Thinning of wood lots to obtain wood, if carefully done, can benefit trees by reducing crowding. Less competition for groundwater results in a healthier forest that is better able to resist insects and disease. Wood burning is also cheaper than conventional fossil fuels in many instances (see sidebar).

Wood burning has a "darker" side, however. Heating a home with wood requires a great deal of work, especially if you cut, haul, and split your own wood. Although hard work is good for the body and the soul, and many people enjoy it, wood heat requires more work than many people are willing to do.

Wood burning can also produce indoor air pollution. Smoke and gases escaping from a wood stove when improperly opened or from leaks and backdraft can pollute indoor air. Soot deposition on walls, ceilings, and drapes is not only a nuisance, it is a sign that the air is polluted with potentially harmful particulates. Wood stoves, even clean-burning models, contribute to outside air pollution.

Wood stoves also tend to produce hot, dry, and uncomfortable interiors. They're rough on sinuses and nasal passages.

Because most models don't come equipped with air circulators, some provisions must be made to move heat throughout the house. Ceiling fans help with this. But even so, wood heat tends to result in hot and cold zones. Peripheral bedrooms, for example, may be cold while the living area bakes.

Because hot air rises, wood heat is problematic in homes with vaulted ceilings. In such instances, hot air rises away from occupants. Second-story rooms or lofts may become uncomfortably hot. Although ceiling fans can help force the warm air back down, you still may need to burn more wood to achieve comfortable temperatures. The more wood you burn, the more it costs, and the more pollution you generate.

Because wood stoves require tending, they are useless as back-up heat when you are away from home. Only the automated pellet stoves (discussed below) will keep a home warm in such instances, so long as the pellet supply in the hopper lasts. As a result, building departments generally will not approve a wood stove as a back-up heat source.

Improperly installed and maintained, wood stoves can also be a fire hazard. Many a house goes up in flames each year as a result of either poor installation or improper stove maintenance.

A final drawback of wood stoves is that they are not as efficient as other forms of back-up heat.

**PARTICULATE EMISSIONS FROM WOOD STOVES**

Older model wood stoves emitted between 30 and 80 grams of particulate matter per hour. The new stoves produce between 3 and 6 grams per hour. That's a reduction of over 90 percent.

**ECONOMICS OF WOOD BURNING**

A cord of wood yields the same amount of useable heat as 200 gallons of heating oil, a ton of hard coal, or about 4,000 kilowatts of electricity. A cord of wood costs about $100 to $150, depending on the location and the type of wood. With home heating oil running $1.50 to $2.00 a gallon, 200 gallons of heating oil will cost about $300-$400. At 8 to 10 cents per kilowatt hour, 4,000 kilowatts of electricity will cost $320-$400. Moreover, oil, electricity generated from coal or nuclear energy, and coal are nonrenewable resources, too. When they're used up, they're gone forever.

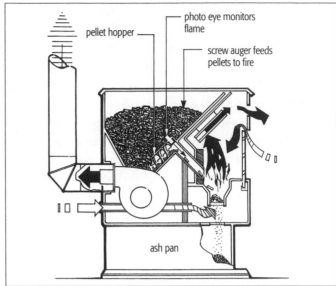

pellet hopper

photo eye monitors
flame

screw auger feeds
pellets to fire

ash pan

**FIGURE 4-21**
*Wood pellets made from
compressed sawdust (a waste
produce of saw mills) are fed
into the combustion chamber
from a hopper.*

## Pellet Stoves

Pellet stoves are the lazy person's wood stove (figure 4-21). Pellet stoves are
fueled by dry, compressed waste wood pellets, made mostly from sawdust, that
in earlier days was often burned at wood mills creating enormous amounts of
air pollution. The pellets are stored in a hopper and fed into the combustion
chamber of the stove automatically by an auger, usually powered by electricity.
Because they are dry and because they're fed into the stove at a controlled rate,
pellets burn very cleanly. Efficient, clean combustion is also ensured by a
steady, well-calibrated flow of air into the combustion chamber. For these reasons, pellet stoves burn as clean or cleaner than most wood stoves. Pellets
come in plastic bags that can be purchased individually or by the ton.

A pellet stove requires electrical energy to operate the auger and the
blower fans that come standard on many models. As in a circulating wood
stoves, the fan increases the rate of heat transfer from the stove to room air.

*Pros and Cons of Pellet Stoves.* Pellet stoves offer many advantages over a
wood stove. They are easy to load, convenient, operate automatically, and
burn a waste material. Many pellet stoves can be thermostatically controlled.
They also allow for more precise control over heat than a wood stove. However, pellet stoves usually cost more than similarly sized wood stoves and
require more service. They require electricity to operate fans and the auger that
feeds the fire. In addition, a pellet stove will generate numerous non-
biodegradable plastic bags each year.

## MASONRY HEATERS

Masonry stoves or heaters are wood-burning stoves made out of bricks and
mortar, rather than welded steel or cast iron (figure 4-22). Designed to burn

much hotter than ordinary wood stoves, they are able to produce more heat energy from wood than a wood stove, and thus make an ideal back-up heat source for passive solar homes.

Masonry heaters burn extremely hot for several reasons. One is that the fireboxes are designed to ensure the ample air flow required for an intensely hot burn. Most designers maximize air turbulence to increase combustion efficiency. Fireboxes are also lined by firebrick or cast panels that insulate the combustion chamber, allowing internal temperatures to reach 1,200 to 2,000°F—hot enough to burn off all liquids and gases. As a result, masonry heaters achieve efficiencies ranging from 88 to 95 percent.

Another distinguishing feature of masonry heaters is the flue. As shown in figure 4-23, in a masonry heater hot flue gases travel through a mazelike channel before escaping. This results in maximum heat transfer, and helps account for the heater's high efficiency.

Masonry heaters also contain a huge amount of mass—up to eight tons in some cases. This mass absorbs the heat produced by the superhot fire. The heat is then slowly radiated into the room, over a period of six to twenty-four hours depending on the mass of the stove after a single two-to-four-hour burn, producing a gentle heat unlike a wood stove. Heat given off by the stove radiates from its surface outward like heat from the glowing embers of a fire, warming the walls, floors, furniture, and people. All solid objects in its "path" are heated. Writing in *SNEWS* magazine, Jay Hensley summed it up best: "Masonry heaters rely on wood's capacity to give off tremendous heat quickly and the ability of the masonry materials to soak up that heat and release it slowly."

**FIGURE 4-22 (left)**
*Masonry heaters are not only functional, many are quite attractive.*

**FIGURE 4-23 (right)**
*One of the keys to the success of a masonry heater is the flue arrangement. Flue gases are forced to take a circuitous route before escaping. Traveling through the mass of the stove, the flue gases give up much of their heat. It is absorbed by the mass and slowly radiated into the room, providing gentle heat that can last for hours.*

courtesy Biofire, Inc.

Nicholas Lyle and Kristin Musnug

In addition to providing back-up heat, masonry heaters can double as thermal mass. Locating a heater in an area that is bathed in sunlight allows the stove to absorb sunlight during the day and radiate heat at night without having to start a fire in it.

Although they're relatively rare in the United States, masonry stoves are growing in popularity. In many parts of Europe, a place where frugality is still held in high esteem, masonry stoves are exceedingly popular. In Finland, for example, 90 percent of all new homes are heated with masonry stoves thanks to generous tax incentives offered by the government. Masonry stoves are also still popular in Norway, Sweden, Denmark, and Germany.

## Pros and Cons of Masonry Heaters

Masonry heaters provide back-up heat for passive solar homes or they can be the main source of heat in solar homes during periods of little sunlight. Masonry heaters can be designed to heat a single room or an entire house of 1,500 to 2,000 square feet, especially if the heater it is strategically located so that its heat distributes evenly throughout the structure.

Masonry heaters are as easy to operate as a wood stove. Most masonry stoves are fired once or twice a day, using 35 to 50 pounds of wood, depending on the stove design for each firing. They burn efficiently and produce very little pollution. A single firing generates an enormous amount of heat that is radiated into a room over a long period, providing hours of gentle comfort.

Like Phase II wood stoves, masonry stoves burn cleanly, so cleanly in fact that they are approved for use when wood-burning bans are in effect. Temp-Cast masonry heaters, one of several commercially available stoves, for example, have been approved by the U.S. Environmental Protection Agency as clean-burning heaters. They have also been approved for use in three of the toughest locations in the United States: Colorado, Washington, and San Luis Obispo County, California. Masonry stoves fall into a super-low-emissions category.

Because masonry heaters burn efficiently, they are safer than wood stoves. Creosote buildup in the chimneys of masonry stoves is not a problem, nor are creosote fires. Masonry heaters are also safer than wood stoves because the surface temperature of the stove is usually no more than 155° to 175°F. In fact, masonry heaters are cool enough to rest against. In Europe, historically, many masonry stoves were built with sitting benches or sleeping platforms. The surface temperature of a wood stove is much higher, casily reaching 400° to 700°F, so that contact with human flesh will result in serious burns. Young children are especially prone to this hazard.

Because the air in a home heated by a masonry stove is not directly heated by the fire, as it is in wood stove-heated homes, indoor air seems cooler and fresher. Because a masonry stove does not superheat the room air, there are no uncomfortable drafts.

Masonry stoves can also be quite visually appealing. Facings of brick, natural stone, rocks, adobe, marble, tile, and stucco can be applied, giving a masonry stove a look that blends nicely with almost any architectural style and decor (figure 4-24). There aren't many heating systems that you'd want to prominently display in your living room. This is one of them!

Further adding to their utility, many masonry stoves can be constructed with built-in bread or pizza ovens. Glass doors can also be installed to view the fire. Some builders use masonry stoves to heat water for radiant-floor heat and hot water for showers and other domestic uses.

Masonry stoves require very little maintenance. An annual chimney cleaning is advised, although probably not necessary. Small amounts of fly ash may need to be removed periodically from the flue. (Accordingly, a masonry stove should have a clean-out door that permits access to the smoke channels.) In central Europe stoves are cleaned once every five to ten years. In *The Book of*

*You wouldn't put your furnace in your living room, but a Finnish stove made out of brick, stone, or tile is a thing of beauty.*

PETER MOORE, MASON

courtesy Maine Woods Wood Heat Company

**FIGURE 4-24**
*A Finnish masonry heater made by Albert Barden III, of Norridgewock, Maine.*

**FIGURE 4-25**
*Tile stoves like this one are ornate but expensive.*

*Masonry Stoves*, David Lyle tells of the Viennese stove builder Gustav Jung who checked his own stove for the first time and found that the flues were still clean. The stove had been built twenty years earlier.

Masonry heaters are not widely available. To contact a mason near you, check the Masonry Heater Association of North America's Web site listed in the Resource Guide at the end of the book. If you can't locate a mason in your area, know that many masons will travel to your site to build a stove for your home. (Masonry stoves are generally built on site.) Vashek Berka of Bohemia International in Lyons, Colorado, sells tile stoves, a type of masonry stove that is adorned with ornate tile. After consulting with a client, he sends details to masons in the Czech Republic, who custom design a stove for the home (figure 4-25). The Czech workers cut the tile and ship it to the United States, accompanied by a skilled mason. Using refractory brick purchased locally, the mason builds the stove on site, usually in about ten days, anywhere in the U.S. The results can be stunning. Or, you can buy a kit and hire an experienced fireplace mason to install it for you. You can even install it yourself, although this option carries with it the danger of making serious mistakes. If you are more daring, you can build your own Finnish fireplace (masonry heater) from scratch with instruction provided by Albert Barden, III (see Resource Guide for his books and address.)

Masonry stoves do have some drawbacks. Like wood stoves, they require an active operator. Because there's a lag time between firing and heat reaching the surface of the stove, masonry stoves are not a quick source of heat. Operators must anticipate their use. Firing mid-to-late afternoon will begin to provide heat for a family within a few hours, depending on the mass of the stove, and will keep a house warm throughout the night and into the next day. If you leave your home for lengthy periods in the winter, you may need to install an automated heating system as a back-up to a masonry heater. This, of course, adds to the cost of building a home.

Masonry stoves must be carefully sized. Generally, the colder the climate, the greater the mass. High-mass stoves provide a steady flow of heat for longer periods than low-mass heaters. In the coldest climates, five-ton stoves would not be uncommon. Bill Eckert sells Temp-Cast stoves and Moberg EPA-certified high-mass fireplaces. The Temp-Cast kits weigh 2,800 pounds. When facing is added, the stoves range from 2 to 4 tons. Bill has seen stoves weighing in at 6 to 8 tons. Lower-mass models tend to cool down more quickly and would require more frequent or longer burns. In less cold climates, lower-mass models work better.

Masonry heaters are fairly expensive. A custom-made stove can cost $11,000 to $15,000 or more, depending on size and details. Building one yourself, which is a chancy proposition, could cost $5,000 or more. Tulikivi stoves, which come assembled, cost between $7,500 and $15,000, depending on the size of the unit (figure 4-26). Their high cost is partly due to the fact that they're imported from Finland. Temp-Cast, which you assemble yourself, is a less expensive stove. It costs about $3,500 but may require a skilled stove mason to apply a facade. As noted earlier, Eckert recommends against installing a Temp-Cast on your own. "Even with a kit," he says, "there are too many places for error." His stoves average $6,000 to $8,000 installed, but can run as high as $8,000 to $10,000 with options like a bread oven and a glass door for viewing the fire.

Masonry heaters must also be carefully built. Because they are so massive, they typically require beefed-up foundations or floor framing to support them. That makes it difficult to add them to existing homes, and increases the cost of a new home. Furthermore, masonry heaters need to be built airtight. Materials must be capable of withstanding high combustion temperatures, and accommodating the expansion and contraction of the mass as it heats and cools. Peter Moore, one of the nation's leading masonry stove builders, notes "Since these stoves go from room temperature to 1,600°F (or higher) inside the firebox, the mason must be aware of how the mass will expand and contract with these temperature swings and plan for it so the stove will not heave itself apart."

When facing material such as tile or brick is installed, it must be spaced properly so it doesn't crack when the unit expands and contracts. Temp-Cast heaters incorporate a "floating firebox" that isolates the heater core from the external masonry facing. This design feature prevents the expansion and contraction of the firebox from damaging the finished exterior.

If not built correctly or from the right materials, cracking can occur. As David Lyle puts it, "heat stress is the Achilles heel of the masonry stove." Mother Nature, Lyle points out, will create an expansion joint where the mason "failed to anticipate the need for one." Further expansion and contraction may eat away at the stove's structural integrity. Cracks also permit pollutants to escape.

Attention must also be paid to their location in a home. Free-standing applications are best. That way, the stove can give off heat on all sides. When placed in corners or against walls, heat radiation is restricted.

Masonry stoves and their chimneys must be placed completely within the warm envelope of a home. That way, the chimneys operate as efficiently as possible. If a chimney is placed against an exterior wall outside the heated space, hot gases may not rise very effectively, especially when a fire is first ignited or after the fire cools down. A tall, warm interior chimney produces the best draft, while a cold exterior chimney causes stubborn lighting, smoky fires, and smoke spillage back into the room.

## MASONRY HEATER KITS

Homeowners and builders can purchase masonry heater kits such as the one sold by Temp-Cast. This heater, made in Toronto, is shipped to building sites on a pallet. Facing stone, brick, stucco, or tile is added to complement the decor. The Moberg Fireplace, a masonry stove manufactured in Portland, Oregon, is also available in kit form. Although you may be inclined to want to assemble your own kit, building a masonry stove is a task only to be undertaken "by talented, careful homeowners," according to Bill Eckert, past president of the Hearth Products Association. In fact, at least one manufacturer (Finland's Tulikivi) produces modular stoves that must be installed by Tulikivi crews. Moberg and Temp-Cast stoves, however, can be assembled and installed by homeowners.

## NO ORDINARY MASON WILL DO

Because masonry stoves require considerable knowledge of materials and how they respond to intense heat, most stoves are built by skilled masons with years of experience in masonry stove building. No ordinary mason will do.

**FIGURE 4-26**
*The Tulikivi masonry heater comes in a kit but can only be assembled and installed by licensed masons.*

*Masonry heater building is an exacting discipline, with much to learn concerning expansion joints, wood-heat technology, venting dynamics, thermal stress and more.*

JAY HENSLEY, SNEWS,
*The Chimney Sweep News*

In airtight homes, make-up air needs to be supplied to masonry stoves. This is another reason it is difficult to retrofit a home for a masonry stove, although it adds a little expense when building a new home.

If make-up air is not supplied to the fire, air can be drawn down the chimney. Known as *backdraft,* this may occur when an exhaust fan is operating. Exhaust fans can create a negative pressure in a house, which causes air to flow into a house any way it can. When a masonry stove has cooled down, chimney temperatures can drop to the point that the draft stalls. Negative pressure can then suck air into the house through this conduit, releasing pollutants into the indoor air.

A damper can also be installed to shut down the flue when the fire goes out to reduce heat loss. Dampers also ensure stronger chimney draft when a fire is first ignited by preventing cold air from entering the chimney. The most effective damper, according to several sources, is a rooftop unit controlled by a stainless steel cable routed through the chimney flue and attached to the stove at a convenient location. The rooftop damper also prevents rain, snow, and wayward birds from invading your home through the chimney, and deflects downdrafts caused by strong winds.

Finally, masonry stoves require more investment in time than conventional heating systems such as radiant-floor heat or baseboard hot-water. Masonry heaters require periodic firing and a knowledge of proper fire construction. Knowing how to build a fire and knowing how to operate a masonry stove are prerequisites for their success and for comfort and safety.

## AIM FOR SUSTAINABILITY

There are many ways to provide back-up heat. As you have seen in this chapter, some options, such as masonry heaters, wood stoves, and solar hot water, use renewable resources for fuel, emit less $CO_2$ and other pollutants, and are therefore more sustainable than others. Even if you choose a more conventional system that is dependent on fossil fuels, for example natural gas, there are ways to make it more environmentally benign, for example, by using a high-efficiency boiler. In doing so, you help make your passively conditioned home even more sustainable.

# PASSIVE COOLING: REGION-SPECIFIC DESIGN

ACCORDING TO THE U.S. Department of Energy, heating and air conditioning account for 44 percent of the energy consumed by U.S. homes. As explained in previous chapters, heating demand can be satisfied in large part through energy-efficient, passive solar design with unmet demand accommodated by a relatively benign back-up heating system. Cooling loads can also be offset by energy-efficient, passive design strategies, an approach known as *passive cooling*.

Passive cooling involves a number of relatively simple, cost-effective measures in the design, construction, and operation of a house. Like passive solar heating, passive cooling achieves comfort without expensive mechanical devices such as air conditioners or swamp coolers, which rely on costly, polluting, utility-supplied electricity.

Although passive cooling can be quite challenging in some climates, with careful design, good construction techniques, and proper selection of materials, it should be possible to cool a home passively in just about any climate—or at least possible to make substantial reductions in a home's annual cooling load.

Passive conditioning can relegate heating and cooling loads to a minor part of total home energy use, greatly reducing the impact of our homes on the environment. Air conditioning an average U.S. home, for instance, consumes more than 2,000 kilowatt-hours of electricity per year. When that electricity is derived from coal, as much of it is, cooling an average home is directly responsible for the release of 3,500 pounds of carbon dioxide and 31 pounds of sulfur dioxide. Passive cooling eliminates these harmful emissions.

As with passive solar heating, the most effective designs and strategies to passively cool a building are determined by the specific climatic characteristics of a given location. Differences in nighttime temperature, atmospheric humidity, wind availability and directional flow, intensity of summer sun, and other factors result in markedly different challenges. Before we examine particular climate-specific design strategies, however, let's explore the basic tools of passive cooling.

## TOOLS AND TECHNIQUES OF PASSIVE COOLING

No matter what climate you are designing for or building in, there are a number of readily available tools to draw on. They fall into four categories: (1) ways to reduce internal heat gain, (2) ways to reduce external heat gains (3) techniques to purge built-up heat from buildings, and (4) methods designed to cool people. A solid understanding of these approaches is vital before plotting a strategy to passively cool a building in a particular climate.

### Reducing Internal Heat Gain

During the cooling season, heat inside homes comes from internal sources such as people, stoves, and appliances, and external sources such as sunlight. One of the easiest and least expensive ways of cooling a conventionally or passively designed home is to reduce internal heat sources to curtail internal heat gain. We will focus on lighting and appliances.

*Lighting.* Among the most common sources of internal heat are incandescent light bulbs. Incandescent light bulbs are highly inefficient sources of light, converting only about 5 to 10 percent of the electricity that flows through them into light. The rest becomes heat. (For this reason, some people suggest calling them heat bulbs!)

Compact fluorescent lights, screw-in fluorescent bulbs that fit into standard lamp sockets, represent an excellent substitute for standard incandescent lighting (figure 5-1). They convert electrical energy into visible light much more efficiently, using 75 percent less energy to produce the same amount of light as an incandescent bulb. Not only do they consume less electrical power, they also emit about 90 percent less heat than a similarly rated incandescent light.

By substituting compact fluorescent light bulbs for incandescent bulbs, especially in areas where lights are on for long periods, a homeowner can achieve marked reductions in electrical use (lower cost and $CO_2$ emissions) and internal heat gain. Lowered heat gain means a cooler interior.

Another way to reduce internal heat gain is through task lighting, a strategy that provides light to high-use or work zones within a room. This technique allows occupants to selectively light portions of a room, rather than bathing an entire room in light when using only a part of the space.

Yet another successful means of reducing heat production by lighting fixtures is daylighting, described earlier as the use of natural light instead of arti-

FIGURE 5-1

*Compact fluorescent light bulbs are color adjusted to provide a light similar to incandescent light bulbs, while using a fraction of the energy of their counterparts. Efficient lighting helps to reduce internal heat gain. Full-spectrum CFLs like this one from Sunwave reportedly create glare-free light that increases visual acuity and reduces eye strain.*

courtesy American Environmental Products

ficial light. In passive solar homes, south-facing windows often contribute significantly to daylighting, as do skylights and windows in other locations. My home rarely requires lighting from sunup to sundown as a result of south-facing glass. To contribute to daylighting without increasing cooling loads, windows in a passive solar home (or a conventional home) need to be positioned so that they permit light to enter during the cooling season without resulting in excess heat gain. That means they need to be placed primarily along the south-facing wall of the home, as explained in chapter 3.

*Appliances.* Household appliances also contribute to interior heat gain. Dishwashers, ovens, clothes dryers, washing machines, and water heaters generate a considerable amount of heat. Although that heat may be welcomed during the heating season, it can cause discomfort during the cooling season. When possible, use heat-generating appliances—which is just about all of them—in the morning or late evening, times during which a little extra heat can be tolerated.

Shifting activities to locations outside the house also helps to minimize internal heat gain. Drying clothes on an outdoor line, for example, and outdoor cooking during the summer months both assist in reducing internal heat gain. Designers and builders can play a role in cutting internal heat gain by providing areas for these activities, such as patios and decks. Shade trees and other vegetation planted around such spaces promote privacy and can help keep these areas cooler as well. A small fountain may also provide additional cooling. By designing outdoor living spaces into a home, an architect not only expands the usable square footage of a domicile, but also fosters an indoor/outdoor existence that many find delightful.

Using less-energy-intensive appliances also helps. Microwave ovens, for example, produce much less waste heat than regular stoves and ovens to achieve the same result, and therefore cut down on internal heat gain. This can be especially helpful in the summer.

New energy-efficient appliances also help to reduce internal heat gain. Because they use less energy, they generate less waste heat than older models. Install energy-efficient models in new homes and replace old, worn-out appliances with newer, more-frugal models.

Finally, in many homes it is possible to seal off heat sources. Laundry rooms and water heaters, for example, can be isolated from the rest of the house with a close of a door. Waste heat can be vented to the outside.

## Reducing External Heat Gain

Important as is internal heat gain, the primary source of unwanted heat in a home during the cooling season is external. External heat gain results in part from sunlight shining on roofs, walls, and windows. Heat produced by sunlight migrates into a home.

External heat gain also occurs as a result warm air surrounding a home. Heat from the air is transferred to roofs, walls, windows, and skylights and then migrates into the interior of a home or may enter in hot air through openings in the building envelope.

Because external heat gain contributes so significantly to cooling loads, finding ways to reduce or eliminate it are vital to passive cooling. Designers rely on a variety of preventative measures, starting with the orientation of a home and window placement.

***Orientation of a Home and Window Placement.*** One of the most significant means of preventing external heat gain and thus cooling a home is proper orientation. As noted in chapter 1, orienting the east-west (long) axis of a passive solar home in the Northern Hemisphere as close to true south as possible minimizes solar gain in the summer. As the orientation of a home deviates from true south, solar gain increases, as does the cooling load. The warmer the climate, the greater the penalty and the more difficult it is to passively cool a home.

***Avoiding Two-story Glass and Skylights.*** External heat gain during the cooling season also results from overglazing. Two-story glass and skylights are particularly troublesome (figure 5-2).

In the two-story passive solar homes I have visited, the upper level of glass is frequently protected by an overhang, but the lower level is not (Figure 5-3). The result of this design flaw is that too much sunlight enters the building from the unshaded lower-tier glazing.

Fortunately, there are ways to shield lower-level glass from intermediate- and higher-angle sun in residences. Cantilevering the top floor, for instance, provides a sufficient overhang (figure 5-4). Horizontal fins also work well (figure 5-5). My advice is to avoid such two-story glass walls.

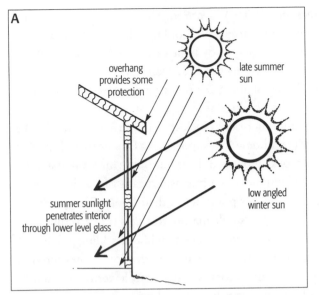

A

overhang provides some protection

late summer sun

low angled winter sun

summer sunlight penetrates interior through lower level glass

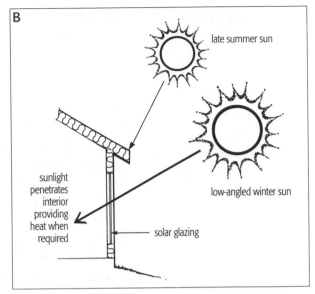

B

late summer sun

low-angled winter sun

sunlight penetrates interior providing heat when required

solar glazing

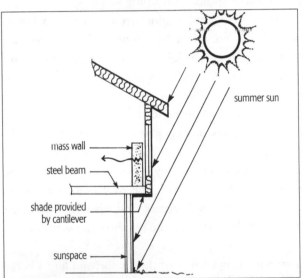

summer sun

mass wall

steel beam

shade provided by cantilever

sunspace

**FIGURE 5-3 (above)**

*In a two-story glass wall (a), the overhang shades the upper tier of solar glazing during the late spring, summer, and early fall, preventing external heat gain. However, sunlight penetrates the lower tier of south-facing glass throughout much of the year because there is no overhang. Compare sunlight penetration of the upper and lower designs.*

**FIGURE 5-4 (left)**

*Cantilevered floor plans help protect lower-level glass from sunlight during the cooling season. Notice the placement of mass in both the upper and lower stories for passive solar heating.*

**FIGURE 5-5**

*In this home, lower-level glass is protected by specially built fins that shade the solar glazing during the late spring, summer, and early fall.*

*Biological Shading: Trees and Other Vegetation.* Shading also reduces external heat gain and is therefore vital to passive cooling. Shade may be provided by vegetation, such as trees and vines, or by built structures, such as overhangs and mechanical shades. According to the U.S. Department of Energy, combined mechanical and biological shading can reduce indoor temperatures by as much as 20°F. Let's examine the biological options first.

Trees and other vegetation not only block the sun, providing shade, they also cool the air surrounding a home by evaporation. This process, known as *transpiration,* draws enormous amounts of heat out of the air around a home. Don't underestimate the effectiveness of a tree as a means of cooling a house. As Anne S. Moffat and Marc Schiler point out in their book, *Energy-Efficient and Environmental Landscaping,* a single mature tree gets rid of as much heat as five 10,000-Btu air conditioners, and it does so without adding pollution to the atmosphere. This is one reason why rural areas with a lot of trees and other vegetation are generally cooler than urban settings with fewer trees. (Another reason is the lack of blacktop and other heat-absorbing surfaces.)

When it comes to providing shade and evaporative cooling for a house, deciduous trees are almost always your best bet. When selecting trees for shade, consider their growth rate, mature height, branch spread, and shape. Plant according to needs. On the east and west sides of a house, remember that the sun strikes from a low angle early and late in the day. To shade walls and windows on the east and west sides, plant bushes or smaller trees. Arbors and trellises also work well here. Vines grow quickly and provide excellent shade.

On the south side of a home, however, it is best to plant high-crowned deciduous trees—that is, trees that grow tall with few low-lying branches. When fully leafed-out in the summer, they shade the roof and south-facing wall. They also permit access to ground-level breezes, and when they lose their leaves in the fall, they permit sunlight to stream into south-facing windows for passive solar gain.

Trees and vines can lower the temperature of the air surrounding a home by as much as 9°F. Grass also helps cool the air around a house, by as much as 10°F. Compared to bare dirt, grass absorbs less sunlight and it loses moisture by transpiration—evaporation from the blades.

Shade can also be achieved by planting evergreens, such as pines and spruces. Be careful, however, where you place them. Because they retain their leaves (needles) year round, evergreens planted along the south side of a passively conditioned home can block wintertime solar gain. Along the north, south, and east sides of a home, evergreens generally work fine, although they may block breezes that help passively cool a home.

*Mechanical Shading Devices.* Window shades and awnings can also help cut down unwanted heat gain, as noted in chapters 1 and 3. Generally, the sunnier the location the greater the need for shading. The most effective shading is provided by external devices: awnings, external louvers, rigid shutters, roller shades, solar screens, and arbors (figure 5-6).

Awnings are fairly inexpensive and relatively easy to install and operate, although they may block views. A properly installed awning can reduce heat gain by up to 65 percent on south-facing windows and 77 percent on east-facing windows. Choose light-colored awnings. They not only shade windows and walls, they reflect sunlight away from a house, reducing heat build-up in the immediate vicinity. (Darker-colored awnings absorb heat.) When installing an awning, be sure to leave a gap between the top of the awning and the side of the house so heat that accumulates under the structure can escape.

External louvers, adjustable slats that control the amount of sunlight entering a window, also provide significant protection against the summer sun (figure 5-7). Slats can be vertical or horizontal and can be adjusted from inside or outside a home.

Shutters are solid or slatted movable wooden or metal coverings that block sunlight entirely and eliminate views when in operation (figure 5-8). Some types of shutters help to insulate windows, providing wintertime benefits as well.

The most expensive external shading device is a roller shutter or shade. Rolling shutters consist of a series of horizontal slats that run along a track. When not in use, they're rolled up out of the way of the window. Roller shades run in tracks, as well, but are made from a sun-resistant fabric (figure 5-9). Both are typically controlled from inside.

Yet another option is a solar screen. Resembling standard window screens, solar screens are mounted internally or externally and thus effectively reduce

**FIGURE 5-6**
*This arbor shades the windows and walls, helping to keep the home cool and comfortable.*

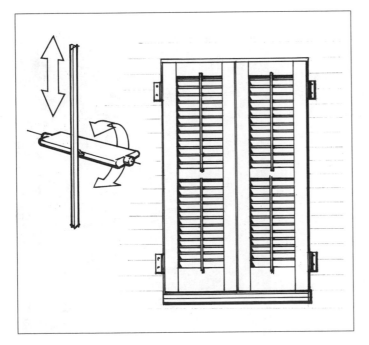

**FIGURE 5-7**
*External adjustable louvered shutters reduce external heat gain yet permit some daylight to enter.*

**FIGURE 5-8**
*External shutters such as these Bermuda-style louvers can block sunlight and views entirely and therefore are not desirable.*

FIGURE 5-9 (top, left)
*Roller shades controlled from
the interior effectively block
sunlight penetration during the
summer and thus help to
reduce external heat gain.*

FIGURE 5-10 (top, right)
*Solar screens block sunlight yet
permit some light to enter.*

sunlight penetration and cut glare without blocking one's view or eliminating
air flow (figure 5-10). Solar screens, like other external shades, provide privacy
and come in a variety of colors and materials.

Interior shades consist of drapes or curtains, conventional pleated shades,
and blinds. Drapes and curtains not only protect against the intrusion of sunlight, they provide decorative value. Tightly woven, light-colored, opaque fabrics work best, as they reflect more incoming sunlight than dark or open
fabrics. Double-layered drapes and curtains block even more sunlight and
improve the insulation value, reducing heat gain during the summer and heat
loss in the winter. For reasons that will be clear shortly, the closer a curtain or
drape is to the window, the better it works.

FIGURE 5-11
*Cellular shades block sunlight
and help insulate during the
summer and winter.*

both photos courtesy Comfortex Window Fashions

Window shades are also effective in blocking sunlight and heat gain. Made from sunlight-resistant fabric, window shades open and close with the pull of a string. Manufacturers produce a wide variety of shades. Some come with a reflective coating to reduce sunlight penetration; others come in dark-colored shades to block incoming light. Still others are honeycombed with two to three air layers to increase R-value, useful in both summer to prevent heat from entering and in winter to prevent heat from flowing out of house (figure 5-11).

Less effective at blocking sunlight than drapes, curtains, and shades are Venetian blinds, metal or wooden blinds with slats that open and close. Although they're less efficient, Venetian blinds do permit some light to enter. Newer models also come with reflective coatings, which increase their effectiveness.

Interior shading devices are generally easier to operate and more convenient than many exterior window shading devices, for example, exterior shutters. They can even be motor-driven and automated. Although motor-driven shades require less operator involvement, they are costly and subject to mechanical breakdown.

Although interior shades offer some significant advantages over most exterior shading devices, they are not as effective at blocking heat gain as external shading devices. This is because exterior shades block sunlight before it penetrates a window; interior shading devices do not (figure 5-12). As a result, considerable heat can build up between a window shade and the glass. Much of this heat will enter the room, increasing the cooling load. So, when selecting a shading device choose wisely.

*House and Roof Color.* Orientation, window placement, and shading combined can reduce cooling loads quite significantly. Further reductions in cooling load occur by painting a house a lighter color. Light-colored walls reflect sunlight rather than absorb it, and thus reduce heat gain. Moreover, light-colored walls increase the lifespan of siding, especially on the south, west, and east facades of a home, which receive the most direct sunlight.

## WHAT'S IN A COLOR?

Darker house colors absorb 70 to 90 percent of the sun's radiant energy. Some of this energy migrates into the house, warming the interior and increasing the cooling load. Light-colored surfaces, in contrast, reflect light away from a home and reduce external heat gain, contributing significantly to passive cooling.

FIGURE 5-12
*External shading devices (a) block sunlight and reduce heat gain more effectively than internal shutters, as shown here (b).*

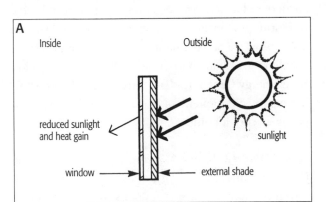

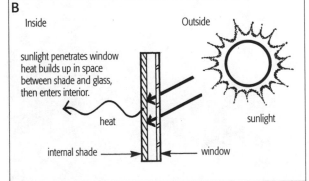

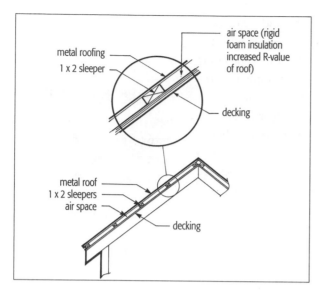

*Metal roofing installed over sleepers creates an air space that reduces external heat gain. Insulation between the sleepers further reduces heat gain during the summer and helps reduce heat loss during the winter.*

FIGURE 5-14

*Spanish tile on these condominiums in Florida creates an airspace that helps reduce external heat gain.*

Heat also penetrates a home through the roof. In fact, most summer heat gain (up to two-thirds) occurs through the roof a home. Changing the color of the shingle or roof tile, however, has little effect on heat absorption. Even white asphalt or fiberglass shingles absorb 70 percent of the sun's radiation. Even so, it is prudent to choose lighter-colored roofs in hot, sunny climates. Although they may not have a huge effect, they will reduce external heat gain. When combined with other measures outlined in this chapter, a light-colored roof and light-colored walls can make a big difference in summer cooling loads.

While roof color may not have a tremendous influence on external heat gain, the type of roofing material and method of installation can dramatically influence it. For instance, metal roofing installed over sleepers creates an air space that retards heat migration into a building, as shown in figure 5-13. Rigid foam insulation placed on the decking between the sleepers further reduces heat gain. Tile roofs also reduce heat gain (especially Spanish tile) because they create an airspace between the tile and roof decking (figure 5-14).

*Radiant Barriers in Roofs.* External heat gain and cooling loads can also be reduced external by installing radiant barriers. A radiant barrier is made from durable aluminum foil that is stapled to roof rafters or attached to roof decking, as shown in Figure 5-15. Radiant barriers block heat penetrating through roofs on hot summer days and are most effective in hot climates. (In cooler climates, heat gain through roofs is less of a problem.) Although the greatest savings come in summer months, radiant barriers also cut down on heat loss in the winter by reducing the flow of heat from the attic to the outside. They are especially helpful in uninsulated spaces such as attached garages. Heat flowing into these areas can easily enter a home, increasing the cooling load.

By reducing heating and cooling loads, radiant barriers reduce energy consumption and utility bills. In Florida, for instance, a homeowner can expect an eight to twelve percent reduction in cooling bills from installing a radiant barrier, according to the Florida Solar Energy Center. Radiant barriers are therefore an important component of an integrated strategy to passively cool a home.

*Good Quality, Low-e Windows.* According to the U.S. Department of Energy, about 40 percent of the unwanted summer heat that enters our homes comes

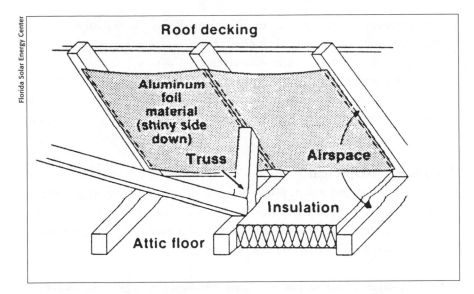

**Roof decking**

Aluminum foil material (shiny side down)

**Truss**

**Airspace**

**Insulation**

**Attic floor**

FIGURE 5-15
*Radiant barriers help reduce external heat gain and are especially useful in hot climates.*

through windows, radiating in from the outside. Because of this, it is important to choose windows wisely. Low-e windows are especially helpful in reducing heat gain and are discussed at length in chapter 2.

*Insulating Your Home.* Although most people mistakenly equate the need for insulation only with heat preservation in cold climates, insulation plays a valuable role in reducing heat gain and keeping homes comfortable during the cooling season.

When used in conjunction with other measures—such as proper orientation and window placement, light-colored exteriors, and shading—insulation helps us achieve cool and comfortable interiors all summer long. Because most external heat gain occurs through the ceiling, pay special attention to roof and attic insulation. For most climates, you will need at least R-30 ceiling or roof insulation, but levels as high as R-50 may be advisable for really hot and really cold climates. Wall insulation is not as important as ceiling insulation for cooling (or heating) because, as just noted, most heat gain occurs through roofs and windows, but don't skimp when it comes to insulating your walls. Floor insulation has little or no effect on cooling, because very little heat enters through floors. However, as noted in chapter 2, floor insulation is important during the heating season in many climates.

*Reducing Air Infiltration.* Hot air entering a home through cracks in the envelope—around poorly sealed doors and windows, for example—also contributes to external heat gain. The contribution varies with the quality of construction.

To prevent this unnecessary problem, passively conditioned homes should be designed and built airtight. As noted in chapter 2, vapor barriers, house wraps, and certain building materials such as insulated concrete forms and structural insulated panels reduce air infiltration. Be sure to seal all remaining cracks in the building envelope with caulk, foam, and weatherstripping.

Controlling air infiltration is inexpensive and pays substantial dividends year round in energy efficiency, comfort, and better air quality, for reasons explained in chapter 6. Reducing air infiltration not only promotes passive cooling, but like many passive design measures, it reduces heat loss during the winter.

## Purging Built-Up Heat

Reducing internal and external heat gains can lower cooling loads significantly. In some climates, these strategies may be all that is required to passively cool a home. In others, additional measures may be needed to remove unwanted heat from internal and external sources.

*Natural Ventilation.* Natural breezes have been used throughout the world for centuries to remove excess heat from buildings, even in hot, muggy climates. On tropical islands, for instance, homes of bamboo permit good air circulation. In some places houses are built on stilts, which further enhances natural air flow. Even today, large modern office buildings sometimes rely in part on natural ventilation to reduce energy costs.

*Cross-ventilation and the Chimney Effect.* In a passively conditioned home, strategically placed windows allow the homeowner to purge heat building up inside a house during the day and the evening by creating cross-ventilation. Cross-ventilation—air flowing naturally from one side of a house to another—can be achieved by opening windows that capture breezes. This works marvelously in many climates, especially if incoming air comes from a shaded area of the yard and other passive cooling techniques are operating. Careful landscaping can facilitate the process. When planted correctly, trees and shrubs can direct breezes toward a house. They can also funnel breezes inward, concentrating or magnify them by the Venturi effect (figure 5-16). Even a gentle, almost lifeless breeze can be concentrated to produce a much stronger flow of air that cools a house at night. When breezes are inadequate, window fans may be required to increase ventilation. Fans can be directed outward to force air out of a house, or inward, to draw air in. A combination of the two may be needed, with incoming air drawn from the cool side of a house and outgoing warm air vented through windows on the warm sides.

Locating a house on a small hill may allow you to tap into breezes not available at other locations on the property. Studying air flow on your property may take some time. Wind currents may vary from season to season. Take your time and plan wisely.

Natural ventilation can also be achieved by the "stack" or "chimney effect," mentioned briefly in chapter 3. Because hot air rises, opening high-

Cross-ventilation is ensured by open floor designs and strategic positioning of openable windows. Naturally occurring breezes facilitate air movement and some windows such as casement windows can actually serve as wind scoops, capturing air flowing around the house and directing it into the interior.

Michael Middleton

**FIGURE 5-16**
*Landscaping and vegetation can be designed to funnel breezes toward a house.*

placed windows in a two-story building permits hot air from lower levels to escape. Replacement air can be siphoned in through basement or first-story windows on the shaded side of a home, cooling the interior naturally. Cupolas can also exhaust hot air from a house.

*Windscoops.* Some designers include windscoops in their homes to facilitate natural ventilation. A windscoop is a tower that captures breezes and funnels them into a house, as illustrated in figure 5-17. Traditionally constructed from high-mass materials such as adobe or other earthen materials, windscoops can be made with conventional building materials, too. Rising high above the roof line, a windscoop captures air flowing 20 to 30 feet above the surface of the ground. This air is slightly cooler than breezes flowing just above ground level, which helps explain why windscoops work so well. However, that's not the only secret to their success. As figure 5-17 shows, air enters the windscoop at the top the structure, then travels down through an underground shaft. As it passes through the ground, it is cooled by the earth before entering the house. In some designs, incoming air is passed through a mist of water or a water-dampened fabric to add humidity. The cooled, humidified air then enters the house, providing a comforting breath of cool, fresh air. Warm air is forced out of the house, escaping through opened windows.

Interestingly, windscoops also help to cool a house when the winds stop blowing through the chimney effect. Sunlight striking the upper portions of the windscoop warms the mass. This warms the air inside the structure. The warm air rises, creating a thermal siphon. This, in turn, draws warm air up and out of the house. Cool air flows in from windows on the shaded side of the structure. During the evening, heat absorbed and retained by the windscoop during the day continues to create an upward draft, drawing warm air out of the

FIGURE 5-17

*Windscoops can contribute to passive cooling but are probably not worth the additional cost if other measures described in the chapter are taken.*

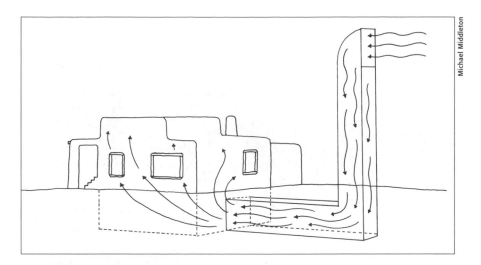

interior. To prevent wind from entering the home when it is not needed or to prevent heat loss during winter months, windscoops have doors that can be opened and closed.

Although windscoops have been used successfully for thousands of years in hot, arid climates, especially the Middle East and Northern Africa, they are rarely used today in new homes. Experience has shown them to be unnecessary, if other passive cooling measures are employed. One homeowner I talked to who lived in a straw bale house in a desert climate found that she relied primarily on superinsulated walls and nighttime purging. Moreover, windscoops have limited usefulness in hot, humid climates. Hot, warm air flowing through a home is not comforting.

*Earth-Cooling Tubes.* Heat build-up can also be relieved by earth-cooling tubes, a technique that received a lot of attention from the building community in the 1970s and early 1980s. Cooling tubes are long, underground pipes that draw outside air into a home, either passively or with fans. The air is cooled naturally by the earth, then delivered to the interior, providing ventilation as well as cool air.

Two types of earth-cooling tube systems exist: open and closed. In an open system, air is drawn into a home, then vented through open windows (figure 5-18a). It follows a one-way or open path through the system. In a closed system, air is drawn into a house, then pumped back through another set of pipes to the earth to discharge heat (figure 5-18b). The cooled air is then circulated back into the house in a closed loop.

Earth-cooling tubes are made from metal and plastic (either PVC or polypropylene). Although both perform well thermally, plastic is easier to install and more resistant to moisture. For optimal air flow, tube diameters ranging from six to eighteen inches seem to work best. Moreover, earth tubes should be buried at least six feet below the ground. In warmer climates, they must be buried deeper, up to twelve feet, to be effective. (The ground is otherwise not cool enough to remove enough heat from the air flowing through earth-cooling tubes to produce comfortable indoor temperatures.)

Although earth-tube cooling seems theoretically sound, designers and builders have very little practical experience with these systems. You are on your own here, so proceed cautiously. Earth-cooling tube systems may require dehumidification, especially in hot, humid climates. Mold can grow inside the cooling tubes. Entrained in the incoming air, mold spores can lead to potentially serious health effects and would surely make a home smell musty. Note, too, that inlets should be carefully protected to prevent insects, rodents, and other animals from entering a home. Finally, earth-cooling tube systems are fairly expensive to install and require high-powered fans to operate.

*Attic and Whole-House Fans.* Ventilating attics reduce heat build-up in houses and can contribute significantly to passive cooling, by reducing heat gain through ceilings. In fact, ventilated attics are about 30°F cooler than unventilated ones.

FIGURE 5-18
*Earth-cooling tubes. (a) Open and (b) closed systems. Air is drawn though the pipes, cooled by the earth, and then enters the home. Fans may be required to facilitate air movement.*

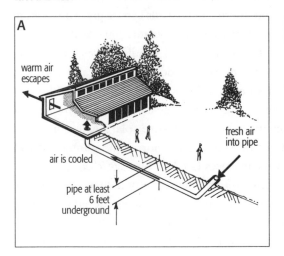

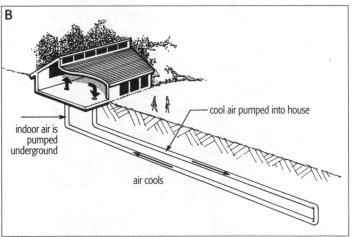

Attics can be ventilated passively or actively. Properly sized and placed louvered openings in the gable eaves of homes, for example, allow hot air to escape passively, as do various types of roof vents. Small fans in roof vents increase the removal of heat and use relatively little power. Even better are solar-powered roof vents, mentioned in chapter 2 (figure 5-19). They are not only effective, they reduce electrical demand. (As noted earlier, installing a radiant barrier reduces external heat gain and attic temperature.)

Far more effective is the whole-house fan. Whole-house fans are relatively easy to install and fairly inexpensive. As shown in figure 5-20, a whole-house fan draws cool outdoor air inside a home through open windows and exhausts hot room air through the attic to the outside

Whole-house fans are typically used to remove heat at night; however, they are typically switched off when outside temperature reaches 85°F the next day. Whole-house fans work well, if regulated correctly, and are much less expensive than central air conditioning systems. However, they are useful only when outside temperature falls below indoor temperature—otherwise, they draw warm air into a home. Unlike air conditioners, whole-house fans can only cool a home to the outside temperature. During the winter, a whole-house fan presents a large, uninsulated hole in the ceiling. If left uncovered, it can leak a significant amount of heat into an attic.

***Thermal Mass.*** Thermal mass can also contribute to passive cooling in some situations. As explained in chapter 3, thermal mass in a house absorbs and radiates heat in response to temperature differences. When the air is warmer than the mass, heat is absorbed. When the air is cooler than the mass, heat is given off. These functions help provide heat in the winter and reduce heat build-up in houses in the summer.

During the cooling season, thermal mass inside a home acts as a heat sink. That is, it absorbs heat from indoor air coming from internal and external sources. Heat gained by tile floors and interior mass walls can be purged from the house at night by natural ventilation—that is, if outside air temperatures fall significantly at night. By simply opening the windows at night and allowing cool air to enter a home, heat absorbed by thermal mass during the day can be eliminated. Even thermal storage walls used for passive solar heating in many climates can be put to use during the summer, as can attached sunspaces (see box on page 182).

Internal thermal mass is especially useful for passive cooling in hot, arid climates (deserts) because nighttime temperatures fall so dramatically, even during the hottest months of the year, due to the lack of

**FIGURE 5-19**

*Attic vents may be used to actively or passively cool attic spaces, thus reducing external heat gain.*

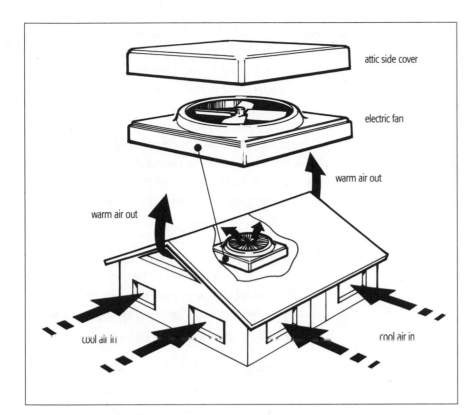

attic side cover

electric fan

warm air out

warm air out

cool air in

cool air in

heat-trapping moisture in the atmosphere. In desert climates, exterior mass walls also help to cool homes (figure 5-21). Exterior mass walls are typically made from concrete, concrete blocks, or earthen materials, such as rammed earth, adobe, or cob. When used to make thick walls, these materials buffer interiors from hot outdoor temperatures. To understand how they work, let's consider an adobe home in the Arizona desert. As the sun peeks above the horizon at dawn, sunlight begins to shine on the exterior walls of the home. This light is converted to heat, which immediately begins to warm the walls. As the day progresses, walls are also warmed as the air surrounding the house heats up. Throughout the day, the walls absorb heat from sunlight and warm air. Heat gained by the mass walls slowly migrates inward, moving down a temperature gradient. By sunset, heat has moved well into the exterior walls but not far enough into them to radiate into the interior, if the walls are built thick enough.

At sunset, the atmosphere begins to cool off and heat absorbed by the walls starts to radiate back into the night air. Heat deep within the wall reverses its flow, moving toward the outside. The wall cools down. By the time the sun rises, the exterior walls are fully cooled and ready to repeat this cycle, which continues day and night, endlessly buffering occupants from intense desert heat.

Important as internal and external thermal mass are in desert climates, in hot, humid climates external mass may be detrimental. The reason is that

## THE NATURAL ALTERNATIVE

Natural building materials such as adobe, cob, rammed earth, cast earth, and rammed earth tires are all excellent choices for desert environments. They can be used to build thick exterior mass walls that buffer interiors from the hot outdoor temperatures.

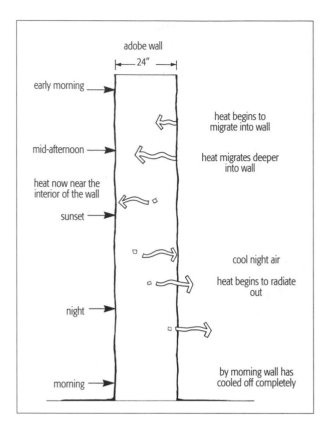

**FIGURE 5-21**
*Exterior walls made from adobe help to regulate interior temperature in desert climates. Note that during the day, heat absorbed by walls slowly migrates inward, but never reaches the interior. At night, heat reverses direction and is released into the nighttime air.*

evening temperatures in such regions often remain too high for nighttime ventilation to effectively purge heat absorbed by mass. However, internal thermal mass can be used in conjunction with active cooling systems that take the place of cool evening air in the deserts. In such instances, rather than open windows at night, the homeowner turns on the air conditioner for a few hours to remove heat absorbed by the mass during the day. Operating the air conditioner at night saves money because the units do not need to work as hard as during the day.

## Cooling People

Reducing internal and external heat gain and purging heat from homes all contribute significantly to passive cooling. However, there is another approach worth considering: cooling people directly.

*Ceiling Fans.* One of the most effective and least expensive means of cooling the occupants of a home is an ordinary fan. Although fans can be used to promote ventilation, which removes heat from a house, they can also be used to move air over and around people, drawing off heat and cooling them.

One of the most popular types is an ordinary ceiling fan with large paddle blades. Ceiling fans can be adjusted to either draw air up from the floor for cooling in the summer or to push warm air down from the ceiling in the winter.

Although portable fans and ceiling fans do not change the internal temperature of a home, they do make us feel cooler. In fact, operating a ceiling fan has the same effect as lowering air temperature by about 4°F. A ceiling fan's cooling effect derives from the fact that moving air strips heat from the boundary layer (a thin region of warm air) around our skin. Removing this warmer air makes us feel cooler. This effect is especially important in hot, muggy climates where evaporation from the skin is reduced by high moisture levels in the air that surrounds our bodies. (In the winter, wind strips heat from the boundary layer resulting in the wind chill, a feeling that outside temperatures are colder than they really are.)

Ceiling fans are effective in all climates, from hot and humid to hot and arid. Although they do require the use of outside energy, fan motors consume very little electricity compared to air conditioners.

Ceiling fans can also be used in conjunction with air conditioning. Researchers at the Florida Solar Energy Center estimate that the use of ceiling fans in conjunction with air conditioners in hot, humid climates allows homeowners to set the thermostat two to six degrees Fahrenheit higher. According to Scott Sklar and Kenneth Sheinkopf, authors of the *Consumer Guide to Solar Energy*, each degree increase saves about 8 percent on cooling costs. It will also substantially reduce noise and environmental pollution caused by the production of electricity.

*Evaporative Cooling and Air Conditioning.* If the measures outlined so far do not provide sufficient comfort, it may be necessary to consider other, more conventional, methods of cooling such as evaporative coolers and air conditioners.

In hot, dry climates, evaporative coolers are effective and relatively energy efficient. An evaporative cooler is a device that draws outside air through a moistened mat. The cool, moist air is then distributed through ductwork or ceiling vents. Evaporative coolers consume much less energy than conventional air conditioners. However, they are of little value in hot, humid climates.

If you can't use an evaporative cooler, you may need to install an energy-efficient air conditioner or a heat pump, which, as described in chapter 4, can provide back-up heat for solar designs and efficient summertime cooling. Be sure to install the most efficient units possible. Place energy-efficient air conditioners in shady locations for optimal performance. Also, be sure to accurately size back-up cooling systems. Undersized systems won't cool adequately and oversized systems are inefficient and wasteful of a homeowner's energy dollars. Oversized equipment, notes Jeannie Legget Sikor, a research engineer at the National Association of Home Builders Research Center and author of *Profit from Building Green*, also dehumidifies air incompletely. "Poor dehumidification occurs because oversized equipment cools the air quickly and does not run long enough to dehumidify." Oversized systems, she adds, also cycle more frequently, increasing the likelihood of failure and thus necessitating more frequent maintenance. If you are including many of the passive cooling features outlined in the chapter, run time should be minimal. (For more details on mechanical cooling systems and listings of performance data, be sure to consult the *Consumer Guide to Home Energy Savings* by Alex Wilson, Jennifer Ihorne, and John Morrill. This book contains a wealth of information, including numerous tables listing the most energy-efficient appliances on the market.

## Combining Measures

Providing comfort in the summer requires intelligent design, an understanding of climate, diligence, and faith—diligence to "design in" a variety of small measures and faith in the cumulative effect of these sensible, cost-effective strategies. Reductions in internal gain through compact fluorescent lights, task lighting, daylighting, and energy-efficient appliances may be small. However, when used in conjunction with techniques that reduce external heat gain such as proper orientation and window placement, shading, light-colored exterior

**BUILDING NOTE**

Several technologies can improve the energy efficiency of cooling systems. While most heat pumps and air conditioners use a single-speed compressor that runs at full speed regardless of cooling needs, some newer units employ variable-speed compressors that run at the capacity appropriate for satisfying the cooling demand. Other systems use variable-speed fans for air distribution that move only as much as necessary to maintain comfort levels and maximize electrical savings.

JEANNIE LEGGET SIKORA,
*Profit from Green Building*

Thermal storage walls, described in detail in chapter 3, are used primarily to heat homes, but they can also provide a bit of cooling. If designed correctly, the mass of a thermal storage wall can be used to draw heat from the inside of the house and transfer it to the outside.

To achieve this effect, you'll need to install vents in the glass, as shown in figure 5-22. When left open on hot summer days, the upper and lower vents exhaust heat that may accumulate in the space between the glass and the thermal mass wall. Combined

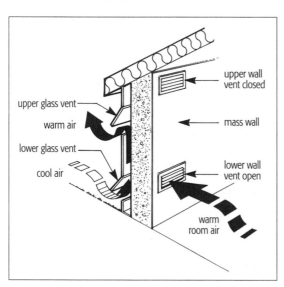

with sufficient overhangs, the vents help to prevent heat from being transmitted into a house.

Venting is especially effective in desert climates. At night, cool desert air circulates through the air space, either naturally or with the assistance of a small fan, drawing away heat that has been absorbed by the mass wall from internal sources such as people, pets, lights, and stoves. By morning, the wall is cooled down and ready for another day of heat absorption.

Some homeowners open the lower interior vents in the mass walls to draw warm room air out of the house at night (figure 5-22). This provides natural

**FIGURE 5-22**
*Vents in the glass portion and the mass of a thermal storage wall contribute to passive cooling in some climates, notably hot, arid ones.*

ventilation and works best if a window can be opened to provide a fresh air inlet. Although thermal storage walls do contribute to passive cooling, their capacity to cool is somewhat limited. If night-time temperatures do not dip below the temperature of the mass wall, as is common in many warmer climates, this strategy isn't very effective. In fact, a thermal storage wall could prove to be a summertime liability unless it is shaded and vented externally.

Attached sunspaces installed primarily for heat gain may also be used to cool a home. Shading, either by deciduous trees or mechanical devices, for example, helps to maintain a cool interior within a sunspace during the cooling season, with spillover benefits to adjacent living space.

Venting is another option for maintaining a cool interior in an attached sunspace and adjacent living space (figure 5-23). As noted in chapter 3, adequately sized vents located in the roof allow hot air to escape passively while low vents placed in the exterior walls allow cool air to enter. For best results,

surfaces, insulation, radiant barriers, and controls on air infiltration, the combined effect can be amazing. Additional gains are achieved by eliminating heat that builds up in a house, for example, by natural ventilation, low-energy active ventilation, and thermal mass. Moving air inside a home with strategically placed fans further advances the goals of natural cooling.

Passive cooling is feasible in any well-built home, such as a standard wood-frame or masonry structure, no matter where it is located. However, many natural building systems, such as adobe and straw bale, are even more

draw air into a sunspace during the summer from cooler areas.

If warm-air vents cannot be incorporated into the roof of the structure, place them on its downwind side. Cool-air-intake vents should be placed low in the walls on the windward side—that is, the side from which prevailing winds tend to flow. This arrangement promotes maximum air exchange and cooling. According to the Sustainable Buildings Industry Council, ventilation area should be at least 15 percent of the total south-facing glass area.

Fans may be used to increase ventilation rates and can be thermostatically controlled to ensure optimal performance of the system. Thermostatic controls are especially important if the sunspace doubles as a greenhouse. Few plants other than cacti and succulants can endure prolonged temperatures above 85°F.

Mass in an attached sunspace can be used to cool a home, too. During the day, mass in the floors and walls of the sunspace absorb heat. If the air cools down significantly at night or if cool night-

time breezes are available, excess heat can be flushed out of the structure through external vents. By morning, the mass is prepared to absorb heat again.

Mass in an attached sunspace can also be used to cool adjacent living space. Mass walls separating a shaded sunspace from living space act as a heat sponge, drawing warmth out of the interior of a home. At night, the heat is released externally by a cold air purge and the mass is recharged and ready to absorb more internally generated heat. Vents in the mass wall can be used to siphon warm air out of the house into a sunspace for elimination.

The key to successful passive cooling of a sunspace is cold nighttime air. Sunspaces in hot, humid climates, where nighttime temperatures in the summer generally linger in the 80s, can't be counted on for cooling. The temperature drop is not great enough to flush much heat from the mass of a sunspace. In fact, mass in the structure may accumulate heat, making adjacent living space and the sunspace extremely uncomfortable.

FIGURE 5-23
*Sunspaces can be vented to provide some passive cooling.*

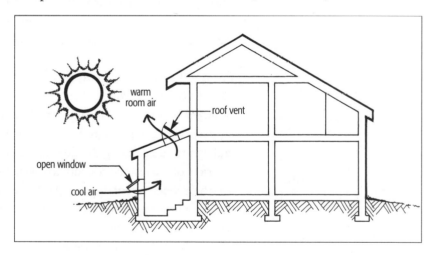

conducive to passive conditioning. The thick walls of a straw bale home coated with plaster provide a superbly insulated shell that reduces heat loss and heat gain, depending on the season, while the plastered surface provides a considerable amount of thermal mass well distributed throughout the house. Earthen walls made from adobe, cob, rammed earth, and rammed earth tires also provide excellent thermal mass, and are ideal for warmer climates (figure 5-25). In colder climates, however, exterior mass walls lose too much heat unless insulated externally.

FIGURE 5-25
*The massive exterior walls of rammed earth tire homes function well in desert climates to ensure passive cooling.*

## REGION-SPECIFIC DESIGN OF PASSIVE COOLING SYSTEMS

Now that you have solid grasp of the tools of passive design, we can turn our attention to their application in different climates (figure 2-26). In this section, we'll examine four climatic zones and outline the strategies for achieving passive cooling in each one. Because many strategies cut across a number of climates, we will focus primarily on the special requirements of each climate. Those who want to learn more about building design in a specific climate should obtain a copy of Joe Lstiburek and Betsy Pettit's *Builder's Guide* for their region (listed in the Resource Guide).

### Passive Cooling in Hot, Arid Climates

Hot, arid climates like the desert southwest or the Australian outback are characterized by relatively short, warm winters during which temperatures rarely fall below freezing. The rest of the year is hot and dry. Keeping a house cool and comfortable during the extended cooling season can achieved by applying a variety of the techniques outlined earlier in the chapter. Most important of all are the proper orientation of the home and proper window allocation and placement (outlined in chapter 3). Overhangs, superior insulation, and radiant barriers are also essential.

Shade is a vital ally in desert climates, too. Unfortunately, the desert is not ideally suited for growing tall shade trees. The lack of moisture means that trees grow slowly. If there are shade trees on the site, protect them!

Shading of walls can also be provided by shrubs planted near the home, vines that drape over trellises or grow against exterior walls, and arbors. Whatever you do, be sure to shade the east and west sides of a desert home. Low-growing trees and shrubs will block the sun early and late in the day.

Awnings and other exterior shading devices may be required to reduce solar gain. Of all your shading choices, external shades are the most effective, as explained earlier in the chapter.

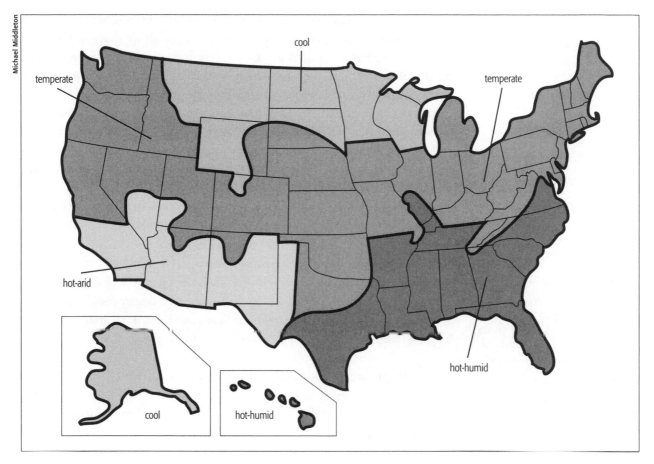

temperate

cool

temperate

hot-arid

hot-humid

cool

hot-humid

Ground vegetation around a home helps to cool a desert home as well. Some cooling results from transpiration, but certain desert plants also reflect large amounts of sunlight and therefore help cool the space around a home.

Indoor plants also help to cool a home in dry desert climates. Houseplants increase the humidity of the air inside a house, making it more comfortable. As Anne Simon Moffat and Marc Schiler point out in their book, *Energy-Efficient and Environmental Landscaping*, houseplants "are extremely effective humidifiers and air coolers and, with a minimum of care, are far more reliable than appliances."

Nighttime purging is another important passive cooling strategy in desert climates. In some desert locations, this is facilitated by cool evening breezes. Vegetation can funnel breezes to a house. If natural ventilation isn't sufficient, you may be able to achieve the same effect with an active system, either window fans or a whole-house fan.

Desert regions can suffer from brutally hot summer winds, too. Vegetation and landforms such as earthberms may be required to deflect hot winds away from buildings. Moffat's and Schiler's book (see above) has recommendations on the types of plants that can be used in desert landscaping to create effective windbreaks.

FIGURE 5-26
*Map of U.S. climate zones.*

Walkways and driveways in desert climates often become unbearably hot as they absorb sunlight during the day. Heat given off by these surfaces can raise the temperature around a house, often dramatically. To create a more comfortable home environment, shade should be provided near driveways, patios, sidewalks, and other paved surfaces by trees and shrubs.

Proper choice of building materials can also play a huge role in maintaining summer comfort. Earthen buildings made from adobe, rammed earth, cob, and even rammed earth tires are ideally suited for desert climates. If your tastes are more conventional, you may want to consider building a concrete home, especially one made from insulated concrete forms.

In life, solving one problem often spawns another. In solving the daytime heating problem in desert climates, for example, by planting trees that grow tall enough to tower over the roof, you may block solar access. If you want to use the sun's energy to make electricity, you may have shot yourself in the foot, unless, of course, you place photovoltaic modules on a pole or a sun-tracking mount system out away from the shade. Proper roof design could also ensure a location for PVs that is free of shade year-round. Earth sheltering can also help enormously in protecting a desert house during the summer.

## Passive Cooling in Hot, Humid Climates

The principal challenge in hot, humid climates is much the same as in the desert: to use design, materials, and landscaping to cool the structure during the long cooling season.

As in desert climates, the first considerations are proper orientation and window placement. Reducing internal and external heat gain are vital. Light-colored exterior walls, overhangs, insulation, and radiant barriers provide effective protection against heat gain on hot summer days. North-facing windows are valuable allies in the battle to stay cool, as they help to dissipate heat.

Although all of these strategies keep a home cool, contending with humidity makes this task much more difficult. Humidity is especially troublesome because it reduces the evaporation from our bodies that cools us down. When atmospheric humidity is high, evaporation slows down. Heat loss decreases and we feel hotter.

Another challenge in such climates stems from the fact that nighttime temperatures remain fairly high. Unlike the desert, in which moisture-free air allows heat to escape into outer space and ambient temperature to plummet, in regions such as the southern United States or the topics, moisture in the atmosphere retains heat. As a result, nighttime temperatures often remain uncomfortably high during the summer. You can't count on relief from the cool nighttime air during the dog days of summer in Alabama.

Windows, roof vents, and skylights can be used to ensure proper ventilation, cooling a house in these climates, especially if it is surrounded by shade trees. They are also effective during the warm but not blistering hot beginning and end of the cooling season. Place openable widows on lower levels or in basements to permit cool outside air from shaded areas to enter a house. The air sweeps through a house, purging it of excess heat, then exits via roof vents, skylights, or high-placed windows. Remember that open designs also allow freer air movement.

If natural ventilation isn't sufficient, you will very likely need to install ceiling fans to create air currents that cool the occupants of a home. A dehumidifier might help increase comfort levels as well.

Covered porches shade a house and provide a cool blanket of air around it. Windows that open onto porches serve as a conduit for cool air to enter a home.

Landscaping can be brought to bear on the problem of summertime cooling. Shade trees cast welcome shadows on the roof, windows, walls, sidewalks, patios, porches, and driveways (figure 5-27). Trees also contribute to a cool blanket of air around the house due to evaporative cooling. The ideal planting consists of tall, high-crowned trees that provide shade but permit breezes to reach the house. In contrast, heavy plantings may block breezes, trapping warm air near a house, making it feel hotter. Although high-crowned trees are desirable, they may block roof-mounted PVs or solar hot water panels. To block sun as it descends in the west each day, plant smaller trees and shrubs.

As in desert climates, cool breezes are a treasured resource in the hot, humid parts of the world and can be funneled to a house through landscaping.

Another design strategy for the landscaper and builder is outside living spaces, such as patios and porches, where you can hang out with friends and family to read, talk, cook, or eat.

In hot, humid climates, the single most important thing you can do to ensure maximum comfort is to protect your home from the midday sun. If you lack funds, get that part of your landscaping underway first. Plant swift-growing trees.

## Passive Cooling in the Temperate Zone

North of the hot southern climates lies the temperate zone. Don't be fooled by the name. The temperate zone is anything but mild. In many regions the winters are cold and windy. Summers may be hot and dry or hot and humid, depending on location. Choose your plan accordingly.

Because the temperate region is a mixed bag of climatic variety, it presents enormous challenges to the house designer and landscaper. The key considerations from the designer's vantage point are proper orientation, proper window placement, overhangs, and superior insulation.

Earth sheltering, either with berms or by burying back walls and the roof, also helps create a home that stays warm in the winter and cool in the summer with little, if any, fossil-fuel energy. The landscaper contributes to comfort by providing summer shade. Deciduous trees planted along the south side of the

house ensure proper shade in the summer months but, after shedding their leaves, permit sunlight to warm the house during the colder parts of the year.

Trees and shrubs may also be needed to shade a house from early-morning and late-day sun to the east and west, respectively. It all depends on the climate in your particular part of the temperate zone. Where I live at 8,000 feet above sea level, summer heat is often a welcome friend. With our cold nights and all of the thermal mass in my house, I find that the interior stays a bit on the cool side in the summer, so I welcome a little extra early morning heat from my east-facing window.

Wind can also present a major design challenge in the temperate zone, especially in flat treeless terrain like that found in the Great Plains of North America. Proper siting can shield a house from fierce, heat-stripping winter winds, as can a carefully planted windbreak, usually consisting of trees. It may be advantageous to funnel cool summer winds toward a house. If summer and winter winds come from different directions, this is fairly easy to achieve. If not, you must choose one or the other. Your decision hinges on the greater good. If winter heating load exceeds summer cooling demand, as it usually does in temperate climates, opt for measures to control winter winds.

## Passive Cooling in Cold Climates

In the cool northern reaches of the United States and other Northern Hemisphere countries, the dominant climate force is cold and the primary goal of designer, builder, and landscaper is to safeguard internal heat during the cooler times of the year.

Summers in the cold climates vary. Some places are fairly cool throughout the summer, so little must be done to shade a house. Others experience searing heat. Overhangs and shade can assist in keeping a house cool. Aside from that, intensive cooling strategies are generally unnecessary.

Earth-sheltered design should be seriously considered in this region. Such designs not only keep homes cooler in the summer, they protect them from wind and help to maintain warmer interior temperatures during the winter by tapping into the Earth's massive reserve of heat. Dehumidification may be required in such instances to prevent mold and mildew during summer months, however. Passive solar design, superinsulation, and a windbreak can greatly reduce your demand for fossil fuel.

## PASSIVE COOLING IS GOOD FOR THE ENVIRONMENT

Applying the principles of passive cooling helps us reduce our dependence on fossil fuels. This strategy also pays several important dividends: It reduces emissions of the greenhouse gas carbon dioxide, which is largely responsible for global warming; it reduces other pollutants associated with the centralized generation of electric power; and it lowers the cost of operating a house during the cooling season. Ultimately, passive solar cooling is good for both people and the planet.

# 6
# HEALTH MATTERS: OPTIMUM AIR QUALITY IN PASSIVELY CONDITIONED HOMES

PASSIVE HEATING and passive cooling are the ultimate in environmentally responsible technology. Supplying comfort to a home or office without noisy, energy-guzzling furnaces or air conditioners, these highly successful approaches to interior climate control are not only gentle on the Earth, they are easy on the pocketbook.

As emphasized earlier, a passively conditioned home must be energy efficient to operate optimally. Making a home energy efficient typically requires a number of measures, as described in chapter 2. Of fundamental importance are measures that reduce air infiltration. However, sealing a home to enhance energy performance can create serious indoor air-quality problems. This chapter examines indoor air quality and offers advice on ways to build energy-efficient, passively conditioned homes that support human health. We'll begin by examining the sources of indoor air pollution in homes.

## INDOOR AIR POLLUTION

Indoor air pollution arises from a variety of sources. Conventional building materials are one common source as are combustion appliances, cleaning products and disinfectants, household pesticides, pets, and even people themselves. The more airtight a home is, the more problems these and other chemicals create.

Many modern building materials contain toxic chemicals that are released into the room air, a process commonly referred to as outgassing. These chemicals can outgas for months, even years, after construction is complete.

Derived from the better-living-through-chemistry movement, most of these products were invented and marketed with good intentions: to make building easier and more efficient. One of the most significant contributors to poor indoor air quality is formaldehyde, a chemical found in many building

**FIGURE 6-1**

*Engineered lumber like oriented strand board used for exterior sheathing, roofs, roof rafters, floors, and floor joists (shown here) is manufactured with a resin containing formaldehyde. This product reduces wood used to build homes. Compare the solid 2x12 (a, right) with the OSB (a, left). Specify low- or no-formaldehyde OSB or air the material out and seal it before use, especially for interior applications such as subflooring.*

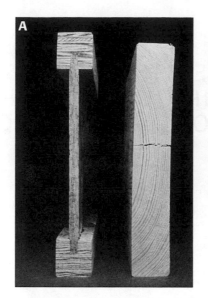

products, including the oriented strand board (OSB) used for external sheathing and subflooring and the particle board used to make cabinets, carpeting, and furniture (figure 6-1). Studies in the 1970s showed that people exposed to low levels of formaldehyde over a long period could become hypersensitized to it. When exposed to even minute concentrations, their health deteriorated markedly.

Although manufacturers have begun to take steps to reduce toxic chemicals in building materials, many new homes still contain a wide assortment of products that release harmful chemicals that can cause illness. In fact, there's hardly a building material that goes into a typical house these days that isn't potentially hazardous to human health. The list of offenders includes carpeting, drapery, vinyl floor tiles, some types of ceiling tiles, upholstery, vinyl wall paper, particle board, plywood, oriented strand board, caulking compounds, paints, stains, and solvents.

Indoor air is also polluted by a host of other sources, including fossil fuel-burning appliances. Stoves, ovens, gas fireplaces, furnaces, and water heaters can all contribute harmful gases such as carbon monoxide to indoor air. Wood stoves and fireplaces also release these gases and potentially harmful particulates into the air inside a home.

Under the kitchen sinks of most homes is yet another source of indoor air pollutants: an arsenal of household cleaning agents and disinfectants. These products contain numerous poisons. Although they may only leak tiny amounts into the air over time from their sealed containers, their true danger occurs when they are put to use scouring pans or cleaning ovens, countertops, and floors. In making homes spic and span, many homeowners are inadvertently releasing harmful pollutants that waft through the air, lingering for hours, sometimes days, poisoning unsuspecting people whose lives they were meant to improve. Ironically, even "personal care" products such as hair sprays and nail polish remover can contaminate indoor air.

Pesticides used indoors to exterminate or repel ants, termites, flies, fleas, moths, and other insects are another source of indoor air pollution. According to a recent survey by the U.S. Environmental Protection Agency (EPA), 75 percent of all Americans used at least one chemical pesticide indoors in the previous year. Although health scientists are concerned about the potential health effects of the active ingredients of these preparations, newer research shows that there is more to fear: Some of the supposedly inert substances found in sprays and liquid pesticide products may be also harmful to human health.

Air fresheners pose the ultimate irony. Some contain the same noxious and harmful chemicals found in moth balls. And don't forget hobbyists sequestered in their basement making duck decoys or model airplanes. Paints and solvents used in these activities emit potentially harmful substances into the air of our homes.

People and pets produce dander (sloughed off dead skin cells) and release potentially harmful chemical substances that may contribute to the deterioration of the air in our homes. In addition, molds, mildew, and bacteria in the air, especially in houses with interior moisture levels above 50 to 60 percent, can cause serious reactions in some people.

Adding to the indoor air-quality problems, carpeting and upholstery often act like sponges, absorbing volatile (easily evaporated) organic compounds from other sources—for example, solvents. Carpets release the contaminants back into the indoor atmosphere for months, even years, ensuring a long and unhealthy exposure.

## THE HEALTH EFFECTS OF INDOOR AIR POLLUTANTS

Indoor air pollutants generally fall into three categories: *toxicants* (toxic substances), *irritants* (substances that irritate), and *allergens* (substances that provoke allergies). Exposure to these substances can have many short-term and long-term health effects.

### Short-Term, Immediate Effects

As with other toxic chemicals, exposure to indoor air pollutants may have immediate health effects, called *acute effects* by toxicologists. These show up soon after exposure to a toxicant, or sometimes after repeated exposure. For example, carbon monoxide is a colorless, odorless gas released from furnaces, stoves, and other combustion sources. At high concentrations, carbon monoxide can be lethal. At lower concentrations, carbon monoxide causes headaches, dizziness, weakness, nausea, confusion, disorientation, and fatigue in otherwise healthy adults and children. These symptoms are often confused with food poisoning or the flu, a misinterpretation that may obscure the real cause

## HOW RADON GETS INTO A HOUSE

- Cracks in solid floors
- Construction joints
- Cracks in walls
- Gaps in suspended floors
- Gaps around service pipes
- Cavities inside walls
- The water supply

Source: EPA, *A Citizen's Guide to Radon*, 2nd ed.

## FORMALDEHYDE-CONTAINING RESINS

Urea-formaldehyde (UF) is used to manufacture particle board and hardwood plywood wall paneling, as well as medium-density fiberboard, which is used to make drawers, cabinets, and furniture tops. Medium-density fiberboard has the highest resin-to-wood ratio of all engineered lumber products.

Phenol-formaldehyde (PF) resins are used to manufacture softwood plywood and oriented strand board. Pressed wood products containing PF resin generally emit formaldehyde at lower rates than products containing urea-formaldehyde resin.

of poisoning for some time. (For this reason, carbon monoxide detectors are now recommended for all homes with combustion appliances.)

Nitrogen dioxide is another combustion gas, produced when nitrogen in air combines with oxygen in a combustion chamber—for example, in a furnace, oven, or water heater. Nitrogen dioxide gas irritates the mucous membranes in the eye, nose, and throat. It may also cause shortness of breath.

## Long-Term, Chronic Effects

Other indoor air pollutants have effects that are manifest years later, typically referred to as *chronic effects*. Radon is an example. Radon is a naturally occurring radioactive gas. This colorless, odorless radioactive pollutant originates in the underlying soil in many parts of the country, seeping in through cracks in basement floors and slabs. The more tightly sealed a home is, the higher the internal levels of radon.

Radon gas is produced during the radioactive decay of naturally occurring uranium in soils, rock, and water. Radon also undergoes radioactive decay, producing small bursts of radiation. The problem with radon gas is not that it emits radioactivity, but that when it decays it produces radioactive lead, a solid. In the lungs, radon gas can decay to lead. Radioactive atoms of lead become lodged in the tissues of the lung where they emit radiation that bombards the cells lining the alveoli or air sacs. This can cause genetic changes in the cells, or mutations, that may lead to cancer, which appears approximately twenty years after the initial exposure. In the United States, physicians annually diagnose an estimated 14,000 cases of lung cancer caused by radon gas, and the majority of the people who contract lung cancer die from it.

Formaldehyde is another prevalent indoor air pollutant. As noted above, formaldehyde is widely used in manufacturing permanent press fabrics, drapes, carpets, fiberglass insulation, paints, shampoos, and plastics. It is also used extensively in resins or adhesives used to produce engineered lumber products such as plywood, oriented strand board, and particle board. For many years, fiberglass insulation was manufactured with a resin containing formaldehyde.

Formaldehyde is released from combustion sources as well. Unvented gas stoves and kerosene space heaters are two common sources of formaldehyde inside homes and work spaces. All in all, it is a pretty difficult to avoid being exposed to this chemical.

Formaldehyde is a clear liquid with a pungent odor at very high concentrations. In room air formaldehyde concentrations are usually quite low, so low that it cannot usually be detected. However, even though you may not be able to smell it, it can still cause health problems, including watering and burning of the eyes and throat, nausea, and difficulty breathing. Some people are sensitive to levels as low as 0.03 parts per million. Higher concentrations may trigger asthma attacks. Long-term exposure may lead to cancer. Although formaldehyde is one of the most widely known indoor air pollutants, it is just one of over one hundred organic compounds present in indoor air.

An unhealthy home has the potential to make its occupants ill, even those who have no history of allergies, chemical sensitivities, or weak immune systems, note David Rousseau and James Wasley in their book *Healthy by Design*. Whether they come from building materials, appliances, cleaning agents, or family pets, indoor air pollutants clearly can have a dramatic effect on human health. No one knows how big the problem is. Asthma afflicts 17 million Americans and 30 percent of all Americans suffer from allergies. According to John and Lynn Bower, authors of *The Healthy House Answer Book*, "these conditions are often caused or aggravated by poor indoor air quality."

## CREATING A HEALTHFUL, ENVIRONMENTALLY FRIENDLY HOME

The goal in building a passively heated and passively cooled home is to create a healthy building that provides the best possible environment for people, including the workers who build them and the families that live in them. In a healthy building, the indoor air is free of toxicants, irritants, and allergens.

To ensure clean air, healthy-house experts recommend a three-part approach: (1) eliminating sources, (2) isolating sources, and (3) ventilation. If these strategies are not sufficient, they typically recommend a fourth action: filtration.

### Eliminating Sources of Indoor Air Pollution

In new construction or passive solar add-ons, the simplest and most effective means of reducing indoor air pollution is to eliminate troublemakers—materials and appliances that produce potentially harmful substances. This strategy, also known as *source reduction*, requires careful product selection—for example, specifying low-formaldehyde or no-formaldehyde oriented strand board or low- or no-VOC (volatile organic compounds) paints, stains, and finishes. Natural building materials such as straw and earth are a good way to achieve this goal, as well. Although there are many types of products a builder should avoid, four deserve close scrutiny.

*Engineered Wood Products.* The first group of products to reduce or eliminate, if possible, are engineered wood products, such as oriented strand board, plywood, particle board, and engineered dimensional lumber. Engineered lumber is made using wood chips, sawdust, thin strips of wood, and other similar material bonded with resins that glue the materials together. Although engineered wood reduces timber cutting, because it makes more efficient use of harvested trees, the resins used to make these products contain formaldehyde.

Engineered lumber products are found in cabinets and furniture. They're used to make wood panels for interior walls, posts and beams, and framing.

*Although a well-built healthy home may sustain or improve the health of its occupants, it is of little value if its construction causes unnecessary damage to the environment. What good is a healthy, energy-efficient home if it undermines the health of the planet?*

**FIGURE 6-2**
*Carpeting is a major source of formaldehyde and accumulates dust, dander, and other potential irritants. Tile and wood floors stained and finished with nontoxic products are good alternatives.*

*Careful selection of heating systems—or better yet, the use of passive solar heating and solar hot water systems for domestic hot water—can substantially reduce the potential sources of indoor air pollution in a home.*

Oriented strand board and plywood are used for subflooring, roof decking, and exterior sheathing.

To create a healthy interior in a new home, avoid manufactured wood products that contain formaldehyde or, if that's not practical, specify low-formaldehyde products. Fortunately, several companies, including Weyerhauser and Louisiana Pacific, now manufacture low-formaldehyde oriented strand board. James M. Huber Company manufactures a high-end no-formaldehyde version. Straw-based particle boards are also available and are made without formaldehyde.

*Carpeting.* The second product to avoid or reduce is carpeting. "Carpeting is a problem," say the Bowers in *The Healthy House Answer Book*, "because it harbors vast quantities of dirt, dust mites, and other allergy-provoking particles." As noted earlier, carpets can become musty; mold and mildew can cause allergic reactions. Dander from people and pets ends up in carpets, too, and can be resuspended in interior room air by vacuum cleaners or other activities.

The Bowers go on to say, "Carpet fibers, padding, chemical treatments, and cleaning products can outgas dozens of harmful chemicals." Most new carpets release formaldehyde (figure 6-2). Volatile organic compounds absorbed last time you painted a pinewood derby car with your children in your basement workshop can be absorbed by carpeting and released over time back into the air of your house.

Minimize or eliminate carpeting. Specify ceramic tile, plaster, hardwood, and wool for wall and floor finishes. Choose environmentally friendly products such as bamboo flooring, natural linoleum, or cork flooring to get the double benefit. They're infinitely healthier.

*Combustion Appliances.* The third area of concern is combustion appliances: wood stoves, fireplaces, gas ranges, furnaces, and water heaters. All can pollute interior air with carbon monoxide and nitrogen dioxide, as noted earlier. Wood-burning appliances may also add particulates to indoor air.

Although eliminating combustion appliances from a house is possible only by substituting the less-efficient and less-cost-effective electric versions, you can install clean-burning or nonpolluting models. Select water heaters, furnaces, and boilers equipped with sealed combustion chambers and outside sources of combustion air to minimize indoor air pollution. Nonpolluting, energy-efficient back-up heating systems are described in chapter 4. Adjusting existing appliances to burn more cleanly also helps eliminate pollutants.

*Nontoxic Paints, Stains, and Finishes.* Other sources of indoor air pollution that should be high on the list of products to avoid when building a passive solar home, a solar addition, or any home, for that matter, are conventional paints, stains, and finishes. Many of these products contain unacceptably high levels of volatile organic compounds (VOCs), including pesticides, mildew-cides, formaldehyde, and solvents that outgas for months after a new home is built or a remodeling job is complete.

Specify low- or no-VOC paints, stains, and finishes to eliminate a significant source of indoor air pollution (figure 6-3). Low-VOC paints, stains, and finishes are now fairly widely available through major retailers and suppliers that specialize in environmentally friendly building products. Be wary, however; according to Cedar Rose Guelberth of the Building for Health Materials Center in Carbondale, Colorado, many of these products still contain small amounts of mercury or lead. To address this inadequacy, some manufacturers such as Wellborn Paint in Denver and Miller Paint Company in Portland, Oregon, produce low- or no-VOC paints made without lead or mercury. They are sold by companies listed in the Resource Guide at the end of the book.

Many environmentally friendly paints also contain chemical fungicides and mildewcides, although the concentrations of these chemicals are relatively small and may pose no health risk. For chemically sensitive individuals or those wary of such chemicals, Safecoat paints and stains produced by AFM Enterprises in San Diego are a good alternative. AFM sells a complete line of people-friendly paints, stains, finishes, sealers, and adhesives. Their products cost more, but contain no formaldehyde, fungicide, and mildewcide, and meet the strictest VOC emissions standards currently in place. BioShield paints are also worth consideration, as they contain no formaldehyde, heavy metals, or biocides.

Casein paints are another benign alternative to conventional wall paints. Casein paints are made from milk protein (casein) mixed with pigments and other substances. The Old-fashioned Milk Paint Company, a family-owned business located in Groton, Massachusetts, manufactures casein paints in sixteen colors. Their paints contain no biocides and are popular among chemically sensitive individuals. Unlike other paints, milk paints are shipped in powder form. Casein paint tends to be flat and streaky, so walls should be coated first with a water-based sealant to prevent water stains.

Most interior and exterior paints these days are water-based. There are some situations, for example on south-facing windows exposed to intense sunlight, where builders prefer to use oil-based paints. When this is the case, select a product that does not contain formaldehyde or such heavy metals as mercury, lead, cadmium, chromium, or their oxides. Select an oil paint that is low in VOCs, too. (VOC levels should not exceed 380 grams per liter). Be sure the paint does not contain any halogenated solvents, either. Suppliers of citrus oil paints that replace standard oil-based paints are listed in the Resource Guide at the end of the book.

Another area of improvement in recent years has been in the production of nontoxic stains, varnishes, and sealers to replace products in wide use today that contain a variety of harmful substances, including potential carcinogens

## ENVIRONMENTALLY FRIENDLY ALTERNATIVES

Environmentally friendly alternatives to the conventional versions of these common products are currently available. Although they are usually more costly, they are worth the additional investment.

Wood primer

Drywall primer

Metal primer

Enamel paints

Latex paints (flat, satin, and high gloss)

Shellac

Lacquer

Wood sealers, including deck sealer

Driveway sealers

Foundation sealers

Concrete, brick, and masonry sealers

Pressure-treated wood

FIGURE 6-3
*Low- and no-VOC (volatile organic compounds) paints, stains, and finishes may cost a bit more, but they're much healthier products.*

and toxicants such as acetone, lead, and pentachlorophenol. Although a number of "healthy" stains and finishes are on the market, they cost more than conventional products sold in most retail outlets. The extra cost is well worth it. They are safer and effective, but also very pleasant to work with. I've used water-based stains and finishes for cabinets and trim and once hired a professional painter to apply some citrus-based finish to the wood in an attached sunspace. He had been in the business for over twenty years and was amazed at how good they smelled and how much better he felt at the end of the day. (For more information on nontoxic building materials, including paints, stains, and finishes, consult the many books listed in the Resource Guide.)

## Isolate Sources of Harmful Pollutants

The second line of defense against toxic compounds in the home is to install barriers to prevent potentially harmful substances from contaminating the air inside a house. This is called the *isolation strategy*.

*Sealing Manufactured Wood.* As noted earlier, engineered lumber such as oriented strand board, I-joists, particle board, and plywood contains resins that contain formaldehyde. If these materials cannot be avoided, be sure to seal them. Oriented strand board used for subflooring, roof decking, or exterior sheathing, for instance, can be coated with nontoxic sealants before being nailed in place. These sealants are available at green-building supply outlets, such as the Building for Health Materials Center in Carbondale, Colorado, or any of the other nearly half-dozen similar mail-order companies in the United States. Cedar Rose Guelberth, who owns the Building for Health Materials Center and consults on healthy building design, recommends AFM's SafeSeal for engineered woods, including oriented strand board and I-joists, and AFM HardSeal for interior applications such as the engineered wood used to make cabinets. Be sure to seal all surfaces: top, bottom, and sides. Water-soluble polyurethane also works well. No matter what engineered wood products you use, be sure to air them out in a dry location before applying the sealant to remove as much formaldehyde as possible. Stack the sheets so that air can circulate over and around all of them. (You can use $1 \times 2$s or $2 \times 2$s as spacers to ensure adequate air circulation.)

Another successful means of isolating potentially harmful substances is to install vapor barriers. Discussed in chapter 2, vapor barriers are typically made of 6-mil plastic and are applied to walls and ceilings. In cold climates, vapor barriers are installed on the warm side of the insulation, that is, on the studs or rafters just beneath the drywall. In warm climates, vapor barriers are installed on the outside of the wall, just beneath the sheathing or decking.

A vapor barrier's main function is to prevent moisture from penetrating the walls and collecting in the insulation, contributing to an unhealthy indoor environment. Moisture reduces the effectiveness of insulation, but may also cause wood and other organic materials, such as cellulose insulation, to decay. In addition to the potential for severe structural damage, moisture accumulating in walls also promotes the growth of mold and mildew. Spores from these microorganisms may enter indoor air through cracks in the wall, causing a whole host of medical problems. Vapor barriers safeguard against these problems, but may also prevent formaldehyde in engineered wood products in exterior sheathing and in insulation from entering indoor air.

*Preventing Radon Gas from Entering a Home.* Radon gas can also be blocked from entering the air inside a home. When building a home, first check out the EPA's map of radon zones. Even if the site is in a region in which radon is not considered a problem, it is wise to test the soil for radon emissions. Most experts recommend testing two locations per building site prior to construction within the footprint of the proposed house. Radon test kits for this purpose, known as radon land test kits, are relatively inexpensive, easy to use, and widely available. I found one manufacturer on the Internet.

If preliminary tests indicate that radon is a problem, you will need to take preventive steps to curb it. Even if preliminary tests turn up negative, you may still have a problem, because levels can vary substantially from one part of the building footprint to another. Radon may be released in only one location under your future home, only a yard or two away from the radon land test device. When the home is built, radon from this one pocket will contaminate the entire house.

Details for preventing radon from entering your home can be found in EPA's *Model Standards and Techniques for Control of Radon in New Residential Buildings,* listed in the Resource guide. Books on healthy building practices also summarize techniques for blocking radon from entering a new building, as well as treating the problem after the fact. A few words here should suffice.

In homes built over crawl spaces, plastic sheeting can be laid over the ground. Be sure to overlap the plastic. Active and passive crawl space ventilation can be used as well, although in cold climates, circulating cold air under a house can make the floors extremely uncomfortable and can waste a lot of energy. Be sure to insulate the floor in such instances. (Actually, you should always insulate a floor over a crawl space!)

In general, basements or slab-on-grade foundations are built on a layer of permeable subslab material, for example, a 4-inch layer of aggregate. A 4-inch porous pipe is laid in the aggregate and is connected to an unperforated vertical riser tube that vents to the outside of the house. This tube passively removes radon from under the slab (figure 6-4), although low-wattage fans can also be used. Over the top of the aggregate, builders often lay a 6-mil layer of plastic sheeting. Overlap the plastic sheeting to prevent radon gas from leaking between adjacent sheets. Be careful not to puncture the plastic when

FIGURE 6-4

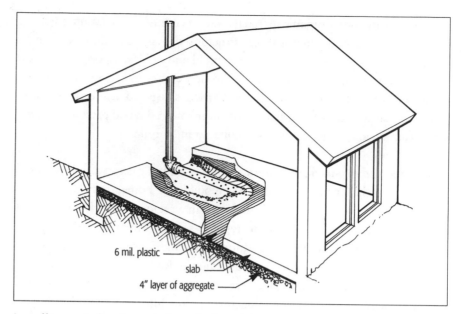

6 mil. plastic
slab
4" layer of aggregate

installing reinforcing steel mesh for slabs and basement floors or when installing radiant-floor heat or pouring concrete. Take measures to minimize cracking of the slab or concrete floor as well.

## Ventilation

The third line of defense is to ventilate—that is, to provide a supply of fresh, clean air to replace the stale, sometimes polluted room air. An energy-efficient solar home should undergo approximately 0.35 to 0.5 air changes per hour. In other words, one-third to one-half of the air in the house should be replaced every hour to ensure healthy indoor air. At this level, additional ventilation is generally not needed, but only if the house has been designed, built, and furnished in accordance with the healthy building guidelines discussed above. In other words, additional ventilation isn't required if a home has been built using nontoxic materials, nonpolluting appliances are installed in the house, furniture and furnishings are free of toxic compounds, and efforts have been taken to seal or block off any potential toxicants.

Making a house even tighter decreases the air change frequency and necessitates additional ventilation, even if one follows all of the rules of healthy building. Natural ventilation is generally the best option. In warm months, opening a window will be sufficient. In cold climates, however, a separate ventilation system will be required to keep the indoor air healthy. We'll examine two basic systems in this chapter, exhaust fans and whole-house ventilation systems.

*Exhaust Fans.* One of the simplest, least expensive means of supplying ventilation is to install (or use) exhaust fans in bathrooms, laundry rooms, and kitchens (figure 6-5).

Located in the ceilings, bathroom exhaust fans force air out of a house. Fresh air comes in through inevitable cracks and openings in the building envelope (figure 6-6). In an airtight, energy-efficient home, bathroom exhaust fans should be used whenever someone is showering or bathing, and whenever the house is closed up, to reduce interior moisture levels and diminish a long string of problems such as the growth of mold and mildew, moisture accumulation in insulation, and water condensation on windows. Exhaust fans can also be run during other times to remove stale or polluted air.

Exhaust fans are typically designed to remove contaminants from the room in which they're installed; they're not generally sufficient to cleanse the air in an entire house. But if you install a timer and let the fan run for several hours a day, it will help provide fresh air. Whatever you do, be sure *not* to divert exhaust into attics or crawl spaces. In attics, warm, moist air can condense and drip down onto insulation, reducing its efficiency and damaging ceilings. Always vent fans outdoors!

Ceiling and kitchen vent fans can be made to work satisfactorily but tend to be rather noisy. Fortunately, there are models that operate fairly quietly, so shop around (see sidebar, page 202).

*Whole House Ventilation Systems.* A far better approach is a whole-house ventilation system, one that exhausts stale, polluted indoor air from one opening in the building envelope and draws fresh replacement air into a house through another. Dedicated whole-house ventilation systems move air through small ducts that transport 50 to 200 cubic feet of air per minute.

To ensure healthful indoor air, as noted above, shoot for ventilation rates of 0.35 air changes per hour or 15 cubic feet per minute per occupant,

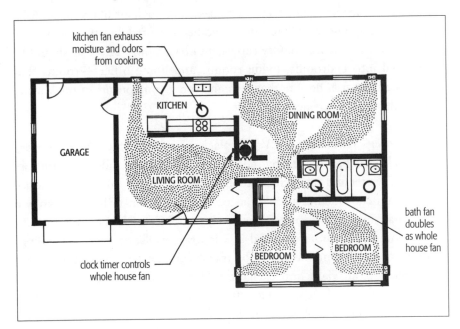

kitchen fan exhauss moisture and odors from cooking

KITCHEN

DINING ROOM

GARAGE

LIVING ROOM

bath fan doubles as whole house fan

BEDROOM

BEDROOM

clock timer controls whole house fan

FIGURE 6-5
*Exhaust fans in kitchens, bathrooms, and laundry rooms can be used to provide whole-house ventilation.*

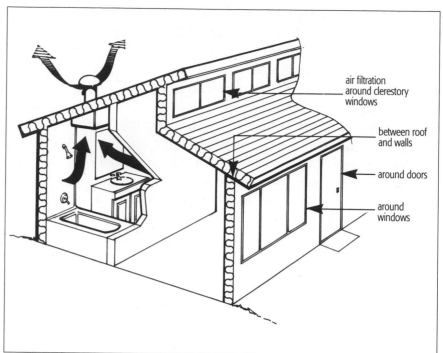

air filtration around clerestory windows

between roof and walls

around doors

around windows

whichever is greater. For a family of four, you will need to ensure a 60-cubic-foot-per-minute ventilation rate. Because fresh air requirements are lower when a house is unoccupied, be sure to install a system that can be adjusted to meet changing needs. You should probably install an oversized unit to accommodate days when your house is brimming with family and friends.

When installing a system, be sure to draw air in from safe locations. Don't place the air intake pipe next to the exhaust pipe of a furnace or clothes dryer. To ensure purity, the incoming air should be passed through air filters (figure 6-7).

In addition, stale or polluted air should be exhausted directly from service rooms, such as kitchens, laundry rooms, and bathrooms. Fresh air should be introduced into bedrooms, living rooms, and home offices—rooms that receive the most use. Be sure that fans and other components of the system, such as filters, are easily accessible, too.

*Heat-Recovery Ventilators (HRVs).* Replacing one-third of a home's air every hour, while essential for healthy indoor air, can cause considerable heat loss during the winter and can increase cooling load during the summer. To reduce energy loss, many builders install heat-recovery ventilators (HRV), also known as air-to-air heat exchangers, in passively conditioned homes (figure 6-8). Heat-recovery ventilators mounted in windows, walls, or basement ceilings conserve energy during the winter by transferring heat in exhaust air to the cool, incoming air via a heat exchanger. During the summer, they ventilate while conserving cool air within a home. In such instances, they transfer heat from incoming air to cool outgoing air.

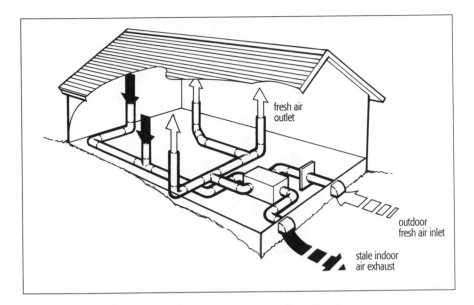

FIGURE 6-7

*A whole-house ventilation system draws fresh air into a house and exhausts stale, polluted indoor air. Air filters can be installed in the system to filter incoming air. This system is integrated with a central forced-air heating system.*

**BUILDING NOTE**

Adequate ventilation equipment is not mandated by most building codes, and is therefore not included in many new homes, unless you request it.

*Integrating a Whole-House Ventilation System.* Whole-house ventilation can be integrated into forced-air central heating and cooling systems, using existing ductwork to distribute fresh air throughout a home. In whole-house ventilation systems, the ventilator fan draws air into the house, which enters the heating and cooling duct system. It is then distributed throughout the house.

Although integrating a ventilation system with a heating and cooling system seems like a good idea, it does have some problems. One of those is that the ducts in forced-air heating and cooling systems are usually very large: typically capable of transporting up to 1,000 cubic feet of air per minute. However, the fans in ventilation systems are rather small and hence incapable of propelling much air through the large duct work. To solve this dilemma, both

FIGURE 6-8

*A heat recovery ventilator is a relatively inexpensive device that helps conserve energy while it provides fresh air to a house and its occupants.*

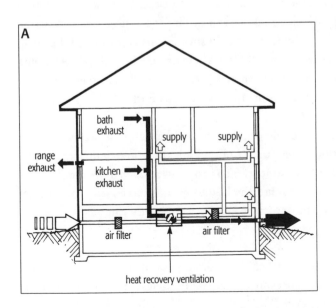

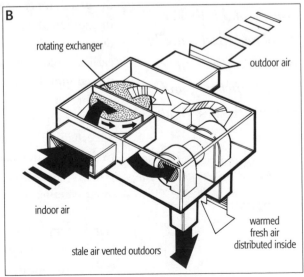

*The secret is not to build looser houses, because loose houses are drafty, uncomfortable, too dry in the winter, and expensive to heat and cool. There are good reasons to build tight houses, and if they're ventilated mechanically, you can have comfort, energy efficiency, and health—the best of all worlds.*

JOHN AND LYNN MARIE BOWER, *The Healthy House Answer Book*

the heating or cooling system and ventilation fans must be operated simultaneously, even when the heating or cooling systems are not operating. Unfortunately, this arrangement generates a lot more noise and requires more electrical energy than an independent whole-house ventilation system.

For more on the subject of ventilation, I recommend the chapter on ventilation in *The Healthy House* by John Bower. After that, you may want to tackle Bower's *Understanding Ventilation*.

*Why Not Build a Leaky House?* It may seem ludicrous to seal up a house, then install a ventilation system to provide fresh air. Why not just build a leaky house and let nature take its course?

The answer to this question boils down to three important concerns: control, comfort, and costs. In a leaky home, owners have little control over the timing and the rate of air movement into and out of a home. On days when the wind is blowing, which often occurs in the winter, houses receive too much fresh air. Drafts result and make the interiors uncomfortable. Moreover, in the winter, a leaky home will require a lot more energy to heat. Uncontrolled ventilation therefore not only renders a house uncomfortable, it comes with a huge price tag: an astronomical heat bill.

On windless days, a leaky house may not receive enough fresh air. That is, there isn't enough air moving into the house to provide adequate ventilation. As a result, pollutants may build up inside the house, reaching unhealthy levels. The air will seem stuffy and will very likely take on a stale odor.

Unregulated ventilation through cracks and crevices also has the potential to permit the transport of unwanted chemicals—such as formaldehyde from insulation and exterior sheathing—into a house. Excess air infiltration therefore reduces your ability to physically isolate yourself from potential toxins.

Excessive leaks and cracks may also drive moisture into the insulation in the walls and ceilings, which could cause a decline it its insulative properties. In addition, moisture may condense on cool surfaces inside walls and then drip down on ceilings, causing an unsightly and costly repair problem. As noted earlier, mold may grow in the moistened insulation and may then be transported into the interior of the home through cracks in the building envelope, causing illness.

As a rule, then, it is generally healthier to build a tightly sealed home with a mechanical ventilation system than to build a leaky home that is vulnerable to the whims of weather. This strategy also produces a much more comfortable house. Bottom line: To create a healthy home you need to control the amount of air entering or leaving the structure, as precisely as you can.

## Air Filters

If a house is constructed with healthy building products, including paints, stains, and finishes, and if measures have been taken to reduce or eliminate internal sources of pollution, to separate occupants from potentially harmful pollutant sources, and to provide adequate ventilation, the inside air should be relatively clean. There are instances, however, when these measures aren't sufficient—for example, when occupants of a home smoke, or when potentially harmful substances such as hair sprays, nail polish remover, pesticides, cleaning products, and solvents are used inside a building. In addition, some people are extremely sensitive to chemicals, even the natural fragrance of pine. Try as you will, you may not be able to eliminate all of the pollutants to which they react.

In such cases, you may need to install an air filtration system. Let me emphasize, however, that air filters are a last resort. In new construction, they are used after pursuing all other options.

When selecting an air filter, it is easy to become confused by details of different models and swayed by misleading sales pitches. Although there are many types of air filters on the market and they fall within a wide price range, there are only two basic air filtration systems: portable and whole-house. Portable filters cost between $80 and $500. Whole-house filters cost a little as $1 and as much as $1,000, including installation.

*Portable Air Filters.* Portable filters purify the air in a single room and work best when run continuously in rooms closed off from the rest of the house—that is, the doors are shut and the air intake and outlet vents for central air or forced-air heating systems are covered (figure 6-9). Otherwise, the volume of air requiring filtration will overwhelm the unit. However, there's a small problem with this strategy: Unless a home is passively heated and passively cooled and each room operates independently, sealing off a room during the heating and cooling season cuts it off from heating and cooling systems.

*Whole-House Filtration Systems.* Whole-house filtration systems are designed to cleanse the air in an entire house. One of the easiest and cheapest ways to install a whole-house filter is to insert a filter or several filters in the duct system or on the heat registers of a forced-air heating system or a central air-conditioning system. That way air circulating through the system will pass through the filters. For the system to work optimally, however, the fan must run several hours each day in the least polluted homes, and more often in more polluted homes (for example, if a home is carpeted, if there is a smoker in the family, or if there are pets bounding about). The system must also run more frequently on "bad air" days or days with high pollen and mold counts.

Fans do not require much electrical energy, and will add only slightly to your monthly utility bill.

*Air filtration is a Band-Aid. Far more effective are strategies to reduce or isolate sources.*

FIGURE 6-9
*Portable air filters like this one made by Ultra Sun Technologies reduce indoor air pollution, but should be used after all other steps have been taken to reduce or eliminate indoor air pollution.*

courtesy Ultra Sun Technologies

However, fans are generally fairly noisy. In addition, you will need a special air filter designed to trap gases and/or particles. Ordinary spun-glass furnace and air-conditioner filters are insufficient, as they're designed to remove only very large particles that could damage a fan motor or clog the coil in a central air conditioner. As such, ordinary filters won't protect people with allergies or asthma.

In houses equipped with a ventilation system that operates in conjunction with a forced-air heating/central air-conditioning system, a single air filter can cleanse fresh air entering the house through the ventilation system and will help remove pollutants from air circulating through the heating and/or central air-conditioning system (figure 6-7).

For passively conditioned homes, a better option is to install an air-filtration unit as part of a whole-house mechanical ventilation system, as discussed earlier. Another option is installation of a heat-recovery ventilator equipped with air filters. This device costs around $1,000 to $2,000.

*Types of Filters.* Filters for portable and whole-house filtration systems come in two varieties: particulate and gas. Particulate filters physically trap suspended particles such as mold, pollen, dust, dander, bacteria, and soot from tobacco smoke or combustion sources such as fireplaces and furnaces. (Note that fireplace and furnace soot is an indication of possible excess carbon monoxide levels and should be corrected at the source.)

*Particulate Filters.* Particulate filters come in three basic types. The most commonly used filters contain fiberglass or other synthetic fibers situated in cardboard or plastic frames. The second type contains a permanently charged plastic film or fiber material that removes particulates from air passing though the filter. The third type consists of a pleated fabric material that traps pollutants.

Particulate filters remove about 10 to 40 percent of the large particulates from room air. Filters for whole-house systems cost anywhere from $1 to $15. Although they're inexpensive, they must be replaced every month or so. In addition, most particulate filters primarily extract large particles, leaving most small and medium-sized particulates in the room air. From a health standpoint, large particulates are less harmful than the very small and medium-sized particulates. Fine particulates penetrate deeply into the lungs, escaping the respiratory system's natural cleansing mechanisms. Heavy metals and toxic pollutants such as insecticides may also adhere to fine particulates. Delivered deep into the delicate lung tissue, these substances can cause serious health problems. Medium-sized particulates are of concern because they can trigger allergies and asthma attacks.

Far more effective are the electronic filters. They can scrub up to 95 percent of the dust, dirt, and smoke—large particulates—from a room. Unfortunately, the terminology and technical details of electronic filtration units can be confusing. However, the concept behind these models is not.

In most electronic units, particulates in room air passing through the filter are given an electrical charge. These charged particles, in turn, stick to oppositely charged surfaces in the room. Particulates removed from room air end up being deposited on walls, ceilings, floors, drapes, tabletops, pets, and people. Although this filter removes particulates very effectively from the air in a home, their accumulation on walls and curtains may be a nuisance. The particles could become re-entrained—that is, they could become airborne again.

Electronic filters such as these, also known as ion generators, may be equipped with a fan and a mechanical filter but are fairly inexpensive, costing between $50 and $150. Because they're portable, you will need to have one in nearly every room of the house or rotate portable filters from room to room.

Another type of electronic filter is the electrostatic precipitator. As in an ion generator, particulates in the air passing through this unit are given a charge. Rather than being deposited in the room, the particulates are captured by the air cleaner itself.

Electrostatic precipators come in portable and whole-house models and are the most expensive particulate filters on the market, running from $150 to over $1,000. One of their main advantages is that they're efficient at removing dirt, dust, and smoke, capturing up to 95 percent of these large particulates. Another advantage is that the collector plates, which trap the charged particles, can be cleaned in a dishwasher or a bath tub, and can be reused indefinitely. There is no need to replace costly filters that end up as landfill. Although these filters remove larger particles efficiently, they only remove about 10 percent of the smaller respirable particles.

*Gas Filters.* Many homes are plagued by a wide assortment of gaseous pollutants, especially volatile organic compounds from newly installed carpeting, paint, furniture, and wood sheathing such as plywood and oriented strand board. These pollutants must be removed by a gas filter. Particulate filters are ineffective in such cases.

Gas filters typically contain an adsorbent material, most commonly activated carbon or activated alumina. These porous substances contain a huge surface area to which gas molecules adhere as they pass through the filter. Activated carbon filters are the most common type on the market and work best on larger volatile organic molecules, such as benzene and acetone. However, these pollutants are not as prevalent in indoor air as formaldehyde. "Treated" activated carbon filters and gas filters containing activated alumina impregnated with potassium permanganate both remove formaldehyde. Note that some manufacturers sell gas filters containing both activated carbon and activated alumina. Note also that although the filter media discussed here are designed to remove gaseous pollutants, they also filter out particulates. However, dust can quickly clog them, so a mechanical particulate filter is usually placed upstream to prevent this problem.

*Hybrid Units.* Because no air filter can remove all indoor air pollutants, many manufacturers sell air cleaners equipped with both gas and particulate filters to ensure maximum protection. Enviracare's portable room filter, for example, contains a particulate filter to remove dust, pollen, and tobacco smoke, and a gas filter to remove numerous pollutants, including cooking odors, paint fumes, and organic compounds (VOCs) emitted by new carpeting and other synthetic materials. The unit sells for about $380 with replacement filters priced at about $60.

Sun Pure Air Purifier, a portable unit, contains six filters (figure 6-9). Air drawn into the unit first passes through a prefilter (mechanical) that removes large particulates. It is then drawn through a second filter, a gas-absorption medium that cleanses the air of formaldehyde and VOCs. Next the air passes through a gas filter containing activated carbon that removes unpleasant household odors and industrial pollutants. This is followed by a HEPA filter (High Efficiency Particulate Accumulator) that removes smaller particles, such as pollen, mold, fungal spores, dust mites, tobacco smoke, and most bacteria—good for asthma and allergy sufferers. The fifth stage is a high-intensity ultraviolet light that kills disease-causing organisms. The final filter is an ionization chamber that produces negative ions, which may help induce and enhance sleep among insomniacs.

The Sun Pure Air Purifier costs $579. Although a unit such as this may sound ideal, a growing number of medical studies over the past two decades suggest that a totally sterile environment may actually be harmful to long-term human health. Much to the surprise of many, this research suggests the strong need for exposure to naturally occurring bacteria, even disease-causing ones, for proper development of a child's immune system. By creating a too-sterile environment, we may actually be causing young immune systems to grow up weaker, which medical researchers believe explains the rapidly rising rate of allergies and asthma among our nation's children.

*Appraising Air Filter Effectiveness.* When buying an air filter you will want to compare models not only by what they remove, but also by their efficiency, how effectively they remove pollutants from room air. The effectiveness of a filter depends on how much air moves through the unit and how well particulate and gaseous pollutants are trapped or removed.

Unfortunately, there are no government standards for reporting air filter efficiency. According to the Asthma and Allergy Foundation of America, "The Food and Drug Administration has twice asked groups of experts to recommend national standards, but neither effort succeeded. Both groups concluded that there isn't enough research data on the relationship between air filtration and actual health improvement to recommend national standards." As a result, buyers must rely on one of two sources of information: (1) manufacturers' claims, which can be confusing or misleading, or (2) efficiency ratings by one of two professional organizations, the American Society of Heating, Refrigeration, and Air Conditioning Engineers (ASHRAE), or the Association of Home Appliance Manufacturers (AHAM). Look for their labels on products to ensure air filtration systems have been tested and to compare competing models. Bear in mind that no rating system is based on health criteria.

ASHRAE has developed a minimum efficiency reporting value (or MERV) for particulate filters. According to Charles Rose, a member of ASHRAE's technical committee on Particulate Air Contaminants and Particulate Contaminant Removal Equipment, "The MERV is a number from 1 to 16 to compare air filters on the basis of the percent of dust they remove from the air. The higher the number the higher the percent."

In February 2000, ASHRAE announced a new standard for testing and rating particulate filters. The test now measures efficiency for a variety of particle sizes, so there can be no misleading claims, such as a manufacturer who claims its filter is 90 percent efficient but fails to tell the consumer that this value refers only to very large particles, which are less of a concern than medium-sized and smaller particles.

Portable air filters are often rated by the Association of Home Appliance Manufacturers (AHAM). Its label on a product lists the clean air delivery rate (CADR), a measure of the effectiveness of particulate filters in removing pollutants such as tobacco smoke, dust, and pollen. According to AHAM, the clean air delivery rate is a measure of the amount of air filtered by a unit, measured in cubic feet per minute, for a specific material. For example, if an air cleaner has a CADR of 380 for tobacco smoke, it reduces levels to the same concentration as would be achieved by adding 380 cubic feet of smoke-free air every minute! Obviously, the higher the CADR, the better the filter.

ASHRAE and AHAM ratings apply only to particulate pollutants, which is by far the most complicated area and the area most rife with faulty or misleading claims. Removal efficiencies for gaseous pollutants are often reported in a more straightforward manner (there is no range of particulate size to deal with), with less room for misrepresentation.

"Although the FDA has no health-related standards," notes the Asthma and Allergy Foundation of America (AAFA), "it does consider some portable air filtration systems to be Class II medical devices." To obtain this rating, a manufacturer must show that the device is safe and that it has a medical benefit. "Look for both the UL (Underwriters Laboratory) seal and a statement of the FDA's Class II approval (on the product). If no FDA statement is available with the device, check the FDA's medical device listing before buying." (Parentheses mine.)

When shopping for an air filter system, the AAFA recommends that you buy a model that can recirculate eight or ten room volumes per hour. "This doesn't guarantee completely clean air," they say, "but it will be much cleaner than with systems that recirculate less." Asthma sufferers should purchase

*The effectiveness of a filter depends on how much air moves through the unit and how well particulate and gaseous pollutants are trapped or removed.*

- House was not built and furnished with healthy materials.
- Precautions were not taken to isolate potentially toxic building materials from room air.
- Combustion sources are not supplied by outside air or vented outside.
- Excess moisture levels evident in the house—for example, condensation appears on windows in cold weather and mold and mildew grow on various surfaces.
- Air is stuffy and smells bad.
- Odors linger.
- Occupants experience ill health after moving into a new home or after remodeling or installation of a new appliance or new furniture and furnishings.
- Someone smokes.
- Toxic household cleaning products, such as hair sprays, and insecticides are frequently used indoors.

systems "that remove more than 90 percent of all particles larger than 0.3 microns in diameter. Most indoor allergens are larger than this, so this efficiency standard will handle them easily." This goal can be obtained by purchasing a unit with a HEPA (high efficiency particulate accumulator) filter.

Many manufacturers now sell HEPA filters with efficiencies over 99 percent. HEPA filters are not only the most efficient particulate filters on the market, they can last up to five years. For most applications, however, they offer more protection than is necessary. The particulates that most bother allergy and asthma sufferers are removed using a medium-efficiency filter. HEPA filters can be used in high-risk areas, such as bedrooms.

Be cautious. There are filters on the market that claim to be HEPAs, but are only half as efficient as a genuine HEPA unit, notes the AAFA. "Insist on a system that meets 'true HEPA' filtration standards. This way you will be certain to get a system that removes at least 90 percent of the indoor allergen particles."

*Does Your House Needs a Filter?* Well-built, well-ventilated homes generally do not require air filters, but how does a builder or a new homeowner know if a house will pass the clean-air test?

Fortunately, there are number of ways to determine whether the air is contaminated. You can hire a professional to test the air or you can rent or buy the equipment and test it yourself, a costly and time-consuming process. Far simpler, you can run through the check list provided in the sidebar. Any of these symptoms indicates the need for ventilation and air filtration.

*Protecting People with Multiple Chemical Sensitivity.* Filtration is highly recommended when one or more of home's occupants suffers from Multiple Chemical Sensitivity (MCS). But even here air filters may create more problems than they solve for people afflicted with this troublesome, often debilitating disorder.

Some air filters generate low levels of pollutants that cause adverse reactions in people with MCS. Fiberglass or polyester fibers in particulate filters, for example, may contain a synthetic resin to which some people are allergic. To enhance their ability to trap particulates, some manufacturers spray particulate filters with a fine layer of oil, which may cause some chemically sensitive individuals to react. In addition, some filters are treated with chemical substances intended to kill mold and other microbes. Although these filters outgas rather tiny amounts of these chemicals at levels that cause no adverse reaction in most of us, they are sufficient to cause reactions in others. One solution is to bake new filters in the oven at 200° F for a couple hours. Check with manufacturers first, because this treatment may cause some filters to deteriorate. Another solution is to install a gas filter "downwind" from a particulate filter.

Another problem with filters, notably the electrostatic precipitators and negative ion generators, is that they produce tiny amounts of ozone. Although most of us aren't bothered by ozone in low concentrations, some people may react badly to it. Another type of particulate filter, the electrostatic type, is made of plastic that outgases pollutants that bother some individuals.

Chemically sensitive individuals may also react to adsorption media in gas filters. Carbon filters made from oxidized coconut husks may be tolerable when carbon from coal is not.

To avoid potential problems, research your options carefully. Talk to suppliers who are aware of the special needs of chemically sensitive people. They can usually set you up with filters that are tolerable to most people. Buy from manufacturers who will allow customers to try a filter in their home and return it if there's a problem.

## PASSIVE SOLAR AND HEALTHY AIR

True integrated design of a passive solar house seeks year-round comfort and a healthy interior. It also pursues a broader goal: protecting the environment from damage. What good is an energy-efficient, passive solar home if it poisons its occupants and wreaks havoc in the planet's life-support systems? When you design and build your home, take the time to consider the strategies discussed in this chapter to reduce or eliminate indoor air pollution.

# DESIGNING A PASSIVELY CONDITIONED HOME AND ASSESSING ITS PERFORMANCE

**DESIGNING A PASSIVELY** conditioned home requires a blend of science, engineering, art, and intuition. Optimal results occur when a designer adopts an integrated or holistic approach, which seeks a beneficial synergy among all building design decisions, materials, and methods. The goal of this design strategy is to ensure optimal comfort day and night throughout the year with the least amount of outside energy, and the least impact on the life-support systems of the planet.

Integrated design reduces the possibility of tragic error in passive design. As I've seen over and over again in my explorations of passive solar, a small error can result in a lifetime of discomfort, higher-than-expected utility bills, and increased environmental impact. How does one go about designing a successful passively conditioned home?

## IMPLEMENTING THE FUNDAMENTALS OF PASSIVE DESIGN

In chapter 1 we explored 14 principles of passive design. They are the starting point in the design of a home. A quick review should prove useful.

To begin, choose a site with good solar exposure. You will need unobstructed access to the sun from 9 or 10 AM to 2 or 3 PM for passive solar heating to work. Next, orient the long axis of the home to the south, within 10° east or west of true south. Remember that rectangular home designs generally work best in a passive design. As you or the designer sketch the home or prepare blueprints, be sure to concentrate windows on the south side of the house, using the recommendations provided in chapter 3. Minimize north, east, and west glazing. Also, be sure to provide overhangs in the building design and shade in the landscape design. If your dream is to achieve greater than 25 or 30 percent of your heat from sunlight, be sure to include additional thermal mass to accommodate greater solar glazing. Position it for optimal function.

Passive design requires a well-insulated building envelope. Use the International Energy Conservation Code as an insulation guideline for ceilings, walls, floors, and foundations. However, because many forms of insulation become much less effective when wet, be sure to design in means to protect insulation from moisture or use a type of insulation that does not decline in R-value when wet. Vapor barriers and ventilation systems both assist in keeping wall and ceiling insulation dry. Ventilation systems have the added benefit of ensuring optimum indoor air quality. Don't forget to specify and use nontoxic, nonpolluting building materials, paints, stains, finishes, and appliances.

As you lay out the floor plan of your home, remember to design the house so that most rooms are heated directly. Open floor plans work well for distributing heat passively. For optimal comfort and functionality, however, be sure to include several sun-free zones for activities such as computer work and reading. Also, design interior space with daily-use patterns and solar gain in mind. Early morning use-zones should be located on the southeast portion of the house.

In addition to adequate insulation, a passively conditioned home requires controls on air infiltration. Seal leaks and cracks to reduce air infiltration, but ensure adequate ventilation for fresh air. If possible, protect your home from winter winds by landscaping, earth sheltering, and vegetation.

When the site is selected and the design for that site is completed, many architects and builders like to estimate the energy performance of the building. This allows them to determine if their goals have been met and to specify back-up heating and cooling systems. To estimate future energy demand, the designer has several options: running the numbers the old-fashioned way, by hand, or turning to one of a handful of sophisticated computer programs to complete this task.

## ANALYZING ENERGY PERFORMANCE BY HAND

Energy performance can be estimated by performing a series of calculations, often presented in worksheet form to facilitate the process. I'll outline the steps in this technique but refer readers to other books on passive solar design, such as *The Passive Solar House* by James Kachadorian, for a more detailed explanation of this process. These books provide the equations, worksheets, reference tables, and extensive explanations of the method. In this discussion, I'll use an example of a passive solar home in New Haven, Connecticut, presented in Kachadorian's *The Passive Solar House*, which is a truly indispensable reference for understanding these concepts and performing these calculations.

To perform an energy analysis of a home by hand, the designer begins by determining the R-value of the building envelope, starting with the walls and roof. This can be determined by adding the R-values of all the components as shown in table 7-1. For an exterior wall, the total includes the R-values of the siding, external sheathing, insulation, drywall, and airspace within the stud space. (Bridging losses, if any, should be deducted from the total for accurate results.) In this example, the walls of the home have an R-value of about 21.4 and the roof has an R-value of about 32.6. Note that most designers and builders know the R-values of standard wall and roof assemblies and can use these numbers for their calculations, rather than adding up each component.

After the R-values of the walls and roof have been determined, the designer must convert them to U-values. You will recall from chapter 2 that the U-value is the inverse of the R-value—that is, U-value is equal to 1/R-value.

The designer uses the U-value and surface area of the roofs and walls to determine heat loss from walls and roofs. After these calculations are completed, the designer must determine the U-values of the windows, along with the surface area they cover, to calculate heat loss through the proposed glazing. Heat loss through foundations should also be calculated.

Because heat is also lost through leaks in the building envelope, the designer must estimate the amount of heat that is required to warm the air flowing into the house through natural infiltration. You will recall that rates of 0.35 to 0.5 air changes per hour are recommended in passively conditioned homes. Formulas needed to make this estimate are presented in *The Passive Solar House*.

The sum of the heat loss through the walls, roofs, windows, and by infiltration is the total heat loss of the building. The result is expressed in Btus per hour (table 7-2).

Now that the designer knows how much heat will be lost through the building envelope, it is time to estimate solar gain—that is, how much heat the home can generate from solar radiation. Solar gain is determined from tables that list solar heat gain factors, the amount of heat produced from sunlight streaming through east-, west-, and south-facing windows (table 7-3). You must use values for your specific latitude.

### TABLE 7–1. Total Wall R-Value

| ITEM | R |
|---|---|
| 15 mph wind (outside) | 0.17 |
| 1-inch rough sawn cedar outside siding | 1.25 |
| 1-inch tongue-&-groove foamboard insulation | 5.00 |
| ½-inch exterior plywood | 0.62 |
| 3½-inch fiberglass batt insulation | 13.00 |
| 6 mil poly | Negligible |
| ½-inch drywall | 0.64 |
| still airspace (inside) | 0.68 |
| Total R-value = | 21.36 |
| Total U-value = 1/21.36 = | 0.0468 |

### Total Roof R-Value

| ITEM | R |
|---|---|
| 15 mph wind (outside) | 0.17 |
| 325# asphalt roof shingles | 0.44 |
| 15# felt paper | 0.06 |
| ½-inch exterior plywood | 0.62 |
| 9-inch fiberglass batt insulation | 30.00 |
| 6 mil poly | Negligible |
| ½-inch drywall | 0.64 |
| still airspace (inside) | 0.68 |
| Total R-value = | 32.61 |
| Total U-value = 1/32.61 = | 0.0307 |

### TABLE 7-2. Total Heat Loss

| ITEM | HEAT LOSS Btus/hr • °F | % OF TOTAL HEAT LOSS |
|---|---|---|
| Walls | 76.14 | 17 |
| Roof | 46.67 | 11 |
| Infiltration | 174.77 | 40 |
| Windows & patio doors | 141.14 | 32 |
| Total = 438.72 Btus/hr • °F | | |

Tables 7-1 through 7-7 reprinted from *The Passive Solar House* by James Kachadorian (Chelsea Green, 1997).

## TABLE 7-3. Solar Heat Gain Factors for 40 Degrees North Latitude

| MONTH | % SUN | DAYS | EAST | SOUTH | WEST |
|-------|-------|------|------|-------|------|
| Sep | 57 | 30 | 787 | 1,344 | 787 |
| Oct | 55 | 31 | 623 | 1,582 | 623 |
| Nov | 46 | 30 | 445 | 1,596 | 445 |
| Dec | 46 | 31 | 374 | 1,114 | 374 |
| Jan | 46 | 31 | 452 | 1,626 | 452 |
| Feb | 55 | 28 | 648 | 1,642 | 648 |
| Mar | 56 | 31 | 832 | 1,388 | 832 |
| Apr | 54 | 30 | 957 | 976 | 957 |
| May | 57 | 31 | 1,024 | 716 | 1,024 |

## TABLE 7-4. Combined SHGF for All Elevations (in millions Btus)

| MONTH | EAST | | SOUTH | | WEST | | TOTAL (millions Btus) |
|-------|------|---|-------|---|------|---|-----------------------|
| Sep | 0.86 | + | 3.72 | + | 0.47 | = | 5.05 |
| Oct | 0.66 | + | 4.37 | + | 0.37 | = | 5.40 |
| Nov | 0.39 | + | 3.57 | + | 0.21 | = | 4.17 |
| Dec | 0.34 | + | 3.58 | + | 0.19 | = | 4.11 |
| Jan | 0.41 | + | 3.75 | + | 0.22 | = | 4.38 |
| Feb | 0.63 | + | 4.09 | + | 0.35 | = | 5.07 |
| Mar | 0.92 | + | 3.90 | + | 0.50 | = | 5.32 |
| Apr | 0.99 | + | 2.56 | + | 0.54 | = | 4.09 |
| May | 1.15 | + | 2.05 | + | 0.63 | = | 3.83 |

## TABLE 7-5. Monthly SHGF Adjusted by Shade Coefficient (in millions Btus)

| MONTH | SC | | MONTHLY TOTAL | | NET TOTAL |
|-------|------|---|---------------|---|-----------|
| Sep | 0.88 | × | 5.05 | = | 4.44 |
| Oct | 0.88 | × | 5.40 | = | 4.75 |
| Nov | 0.88 | × | 4.17 | = | 3.67 |
| Dec | 0.88 | × | 4.11 | = | 3.62 |
| Jan | 0.88 | × | 4.38 | = | 3.85 |
| Feb | 0.88 | × | 5.07 | = | 4.46 |
| Mar | 0.88 | × | 5.32 | = | 4.68 |
| Apr | 0.88 | × | 4.09 | = | 3.60 |
| May | 0.88 | × | 3.83 | = | 3.37 |

Solar heat gain factors from the tables are used to determine the number of Btus of heat each window produces by multiplying the solar heat gain factor by the number of square feet of glass by the number of days in the month. This figure is then multiplied by the percentage of sunshine. These values are entered into a table and added up to estimate solar heat gain for each month of the year for all east-, west-, and south-facing windows (table 7-4). Heat gain is then adjusted (reduced) to compensate for shade coefficient (SC), light reflected off the glass. The final result is a monthly heat gain (table 7-5).

Next the designer must calculate heat demand by month: how much heat the home will require each month of the year. This is determined by multiplying the total heat loss (determined earlier) by the degree days for each month, numbers that can be derived from reference tables (table 7-6). (Tables with values for Solar Heat Gain Factors, percent sunshine, and degree days can be found in *The Passive Solar House*.)

Now that you know the monthly heat demands (heat load) and the monthly solar gain, you can calculate the difference on a month-by-month basis to see how much heat a home can acquire from solar energy. As shown in table 7-7, the difference between heat load and solar gain is the amount of heat that must be supplied by a back-up heating system in an average year.

Monthly heat loads and solar gain values are added to determine the estimated total annual heat load and annual solar gain. These values will determine the percentage of heating supplied by solar energy. In the example shown in the accompanying tables, the heat load is 65 million Btus and the heat supplied by sunlight is 33.7 million Btus. Dividing the solar gain by the heat load indicates that this home would acquire nearly 52 percent of its heat from sunlight.

If a designer is unsatisfied with the projected performance, he or she must start over, adding insulation, more south-facing windows, and other details to reduce heat loss through the building envelope and increase solar gain. The analysis begins again. If, after running the new numbers, the home still performs below expectations, the designer must make additional modifications, and run the numbers over again. Similar calculations need to be performed to determine the performance of a home during the cooling season.

## ANALYZING PERFORMANCE USING WORKSHEETS AND COMPUTER SOFTWARE

Analyzing the performance of a passive design by hand, even with worksheets, is cumbersome, tedious, and time consuming, with an ever-present chance for error. Fortunately, a faster and more accurate way to estimate the energy performance of a passively designed building is available: energy analysis software. Several programs permit detailed energy projections of designs, taking into account numerous factors that influence a building's performance (see sidebar on page 218). Using a computer program dispenses with the need to perform the arduous, complicated calculations to determine a building's performance.

These programs are not only faster, they also facilitate the integrated approach. Designing a home without them is nearly inconceivable to many architects. When asked for recommendations on passive design, NREL's Ron Judkoff said, "Use a building energy simulation model to design passive buildings." These programs run on an ordinary home computer "in a blink of an eye."

Software available for energy analysis of passively conditioned buildings permits three vital functions: (1) analysis of designs for year-round energy performance, both heating and cooling; (2) analysis of design modifications to achieve the lowest possible energy use in residential and commercial structures; and (3) region-specific design. Most programs permit analysis early on, beginning in the concept phase.

In this chapter, I will review several building analysis software packages. Remember that these programs are not intended to be used to draw blueprints, only to analyze designs for energy performance or figure out design details that can be translated into blueprints. Readers interested in exploring additional software tools should consult the Tools Directory, an on-line listing of over

### TABLE 7-6. Household Monthly Heat Load

| MONTH | HEAT LOSS OF HOME | | DEGREE DAY | | MONTHLY HEAT LOSS (millions Btus) |
|---|---|---|---|---|---|
| Sep | 10,529* | × | 117 ** | = | 1.23 |
| Oct | " | × | 394 | = | 4.15 |
| Nov | " | × | 714 | = | 7.52 |
| Dec | " | × | 1,101 | = | 11.59 |
| Jan | " | × | 1,190 | = | 12.53 |
| Feb | " | × | 1,042 | = | 10.97 |
| Mar | " | × | 908 | = | 9.56 |
| Apr | " | × | 519 | = | 5.46 |
| May | " | × | 205 | = | 2.16 |
| | | | Total | = | 65.17 |

* 438.72 Btus/hr • °F × 24 hrs/day = 10,529 Btus/°F • day
** See appendix 4

### TABLE 7-7. Performance Summary (in millions of Btus)

| MONTH | HEAT LOAD | SOLAR SUPPLIED | DIFFERENCE: NOT SOLAR SUPPLIED |
|---|---|---|---|
| Sep | 1.23 | 4.44 | 0 |
| Oct | 4.15 | 4.75 | 0 |
| Nov | 7.52 | 3.67 | 3.84 |
| Dec | 11.59 | 3.62 | 7.97 |
| Jan | 12.53 | 3.85 | 8.68 |
| Feb | 10.97 | 4.46 | 6.51 |
| Mar | 9.56 | 4.68 | 4.88 |
| Apr | 5.46 | 3.60 | 1.85 |
| May | 2.16 | 3.37 | 0 |
| | Total = 65.17 | | Total = 33.73 |

Building energy software falls into four categories:

1. Whole-building analysis
2. Codes and standards
3. Materials, components, equipment, and systems
4. Miscellaneous applications, such as energy economics, air pollution, and water conservation

For a complete listing of computer programs for homeowners, commercial building owners and operators, building designers and builders, and researchers in the United States, Canada, and other countries visit the Department of Energy's Office of Building Technology Web site at www.eren.doe.gov/buildings/tools_directory/

two hundred energy-related software tools for building with an emphasis on renewable energy, energy efficiency, and other topics related to sustainability (see sidebar, left).

## BuilderGuide

BuilderGuide is an energy analysis program available from the Sustainable Buildings Industry Council (SBIC) in Washington, D.C. It is available in Windows or DOS format—but no Macintosh version—and comes with a book entitled *Passive Solar Design Strategies*. The software and the book are customized to your building location, meaning both contain data needed to perform energy analysis at your particular location.

Parts I and II of *Passive Solar Design Strategies* provide a brief introduction to passive solar energy, covering such topics as energy conservation, thermal mass, orientation, and south-facing glass. Part III provides strategies for passive design *in your area*. This section includes information on sun-tempered design, direct gain, attached sunspaces, and thermal storage walls and will help a designer determine glazing requirements, mass, and overhangs, among other things.

Part IV of *Passive Solar Design Strategies* is a set of four worksheets that walk the designer through the energy analysis of a design, providing virtually all of the necessary data.

The worksheets provided in *Passive Solar Design Strategies* permit the user to analyze energy performance at any stage in the development of a design. Starting this process at the beginning permits the designer an opportunity to fine-tune a design to reach the building's performance goals.

Worksheet I, shown in figure 7-1, is used to calculate the Conservation Performance Level—that is, the annual heating load. The annual heating load is the amount of heat a home will require for comfort in a specific region and is determined by degree heating days and heat loss (via conduction and infiltration) through the building envelope and foundation. Heating load is expressed in Btus per square foot per year.

The conservation performance level of a home is then compared to a base case, a traditional home built with less attention to energy efficiency. As an example, a 1,500-square-foot stick-frame passive solar home in Grand Junction, Colorado, needs about 27,000 Btus per square foot per year, while a standard home of identical size would require about 43,000 Btus per square foot per year.

Worksheet II in *Passive Solar Design Strategies*, shown in figure 7-2, is used to calculate the Auxiliary Heat Performance Level—that is, how much heat a home will acquire from passive solar design features and how much will need to be supplied by a back-up heating system.

Worksheet III, shown in figure 7-3, is used to calculate the expected temperature swing in the house without the back-up heating system operating.

## General Project Information

Project Name _____     Floor Area _____

Location _____     Date_____

Designer_____     _____

## Worksheet I: Conservation Performance Level

### A. Envelope Heat Loss

| CONSTRUCTION DESCRIPTION | AREA | R-VALUE (TABLE A) | HEAT LOSS |
|---|---|---|---|
| Ceilings/roofs_____ | _____ ÷ | _____ = | _____ |
| _____ | _____ ÷ | _____ = | _____ |
| Walls _____ | _____ ÷ | _____ = | _____ |
| _____ | _____ ÷ | _____ = | _____ |
| Insulated Floors_____ | _____ ÷ | _____ = | _____ |
| _____ | _____ ÷ | _____ = | _____ |
| Non-solar Glazing_____ | _____ ÷ | _____ = | _____ |
| _____ | _____ ÷ | _____ = | _____ |
| Doors _____ | _____ ÷ | _____ = | _____ |
| _____ | _____ ÷ | _____ = | _____ |

_____ Btu/°F–h

TOTAL

### B. Foundation Perimeter Heat Loss

| DESCRIPTION | PERIMETER | HEAT LOSS FACTOR (TABLE B) | HEAT LOSS |
|---|---|---|---|
| Slabs-on-Grade_____ | _____ × | _____ = | _____ |
| Heated Basements_____ | _____ × | _____ = | _____ |
| Unheated Basements_____ | _____ × | _____ = | _____ |
| Perimeter Insulated Crawlspaces_____ | _____ × | _____ = | _____ |

_____ Btu/°F–h

TOTAL

**C. Infiltration Heat Loss**   _____ × _____ × .018 = _____ Btu/°F–h

BUILDING VOLUME    AIR CHANGES PER HOUR

**D. Total Heat Loss per Square Foot**   24 × _____ ÷ _____ = _____ Btu/DD–sf

TOTAL HEAT LOSS (A+B+C)    FLOOR AREA

**E. Conservation Performance Level**   _____ × _____ × _____ = _____ Btu/yr–sf

TOTAL HEAT LOSS PER SQUARE FOOT    HEATING DEGREE DAYS (TABLE C)    HEATING DEGREE DAY MULTIPLIER (TABLE C)

**F. Comparison Conservation Performance** (from previous calculation or from Table D)   _____ Btu/yr–sf

**Compare Line E to Line F**

FIGURE 7-1

*This worksheet provided in* Passive Solar Design Strategies *with appropriate background information, such as R-values for walls and heat loss through foundations, is used to estimate heat loss through the building envelope.*

## Worksheet II: Auxiliary Heat Performance Level

### A. Projected Area of Passive Solar Glazing

| SOLAR SYSTEM REFERENCE CODE | | ROUGH FRAME AREA | | NET AREA FACTOR | | ADJUSTMENT FACTOR (TABLE E) | | PROJECTED AREA |
|---|---|---|---|---|---|---|---|---|
| _____ | | _____ | × | 0.80 | × | _____ | = | _____ |
| _____ | | _____ | × | 0.80 | × | _____ | = | _____ |
| _____ | | _____ | × | 0.80 | × | _____ | = | _____ |
| _____ | | _____ | × | 0.80 | × | _____ | = | _____ |
| _____ | | _____ | × | 0.80 | × | _____ | = | _____ |
| _____ | | _____ | × | 0.80 | × | _____ | = | _____ |
| _____ | | _____ | × | 0.80 | × | _____ | = | _____ sf |

TOTAL AREA          TOTAL PROJECTED AREA

### B. Load Collector Ratio

$$24 \times \underset{\substack{\text{TOTAL} \\ \text{HEAT LOSS} \\ \text{(WORKSHEET I)}}}{\rule{2cm}{0.4pt}} \div \underset{\substack{\text{TOTAL} \\ \text{PROJECTED} \\ \text{AREA}}}{\rule{2cm}{0.4pt}} = \rule{2cm}{0.4pt}$$

### C. Solar Savings Fraction

| SOLAR SYSTEM REFERENCE CODE | PROJECTED AREA | | SYSTEM SOLAR SAVINGS FRACTION (TABLE F) | | |
|---|---|---|---|---|---|
| _____ | _____ | × | _____ | = | _____ |
| _____ | _____ | × | _____ | = | _____ |
| _____ | _____ | × | _____ | = | _____ |
| _____ | _____ | × | _____ | = | _____ |
| _____ | _____ | × | _____ | = | _____ |
| _____ | _____ | × | _____ | = | _____ |
| _____ | _____ | × | _____ | = | _____ |

$$\underset{\text{TOTAL}}{\rule{2cm}{0.4pt}} \div \underset{\substack{\text{TOTAL} \\ \text{PROJECTED} \\ \text{AREA}}}{\rule{2cm}{0.4pt}} = \underset{\substack{\text{SOLAR} \\ \text{SAVINGS} \\ \text{FRACTION}}}{\rule{2cm}{0.4pt}}$$

### D. Auxiliary Heat Performance Level

$$\left[1 - \underset{\substack{\text{SOLAR} \\ \text{SAVINGS} \\ \text{FRACTION}}}{\rule{2cm}{0.4pt}}\right] \div \underset{\substack{\text{CONSERVATION} \\ \text{PERFORMANCE} \\ \text{LEVEL (WORKSHEET I,} \\ \text{STEP E)}}}{\rule{2cm}{0.4pt}} = \rule{2cm}{0.4pt} \text{ Btu/yr–sf}$$

### E. Comparative Auxiliary Heat Performance (From previous Calculation or from Table G)

_____ Btu/yr–sf

**Compare Line D to Line E**

FIGURE 7-2
*This worksheet is used to calculate estimated solar gain
and back-up heating requirements.*

# Worksheet III: Thermal Mass/Comfort

## A. Heat Capacity of Sheetrock and Interior Furnishings

|  | FLOOR AREA | HEAT CAPACITY | UNIT HEAT CAPACITY | TOTAL |
|---|---|---|---|---|
| Rooms with Direct Gain _____ | _____ × | 4.7 | = _____ |  |
| Spaces Connected to Direct Gain Spaces _____ | _____ × | 4.5 | = _____ | Btu/°F |
|  |  |  | TOTAL |  |

## B. Heat Capacity of Mass Surfaces Enclosing Direct Gain Spaces

| MASS DESCRIPTION (INCLUDE THICKNESS) | AREA | UNIT HEAT CAPACITY (TABLE H) | TOTAL HEAT CAPACITY |
|---|---|---|---|
| Trombe Walls _____ | _____ × | 8.8 | = _____ |
| Water Walls _____ | _____ × | 10.4 | = _____ |
| Exposed Slab in Sun_____ | _____ × | 13.4 | = _____ |
| Exposed Slab Not in Sun_____ | _____ × | 1.8 | = _____ |
| _____ | _____ × | _____ | = _____ |
| _____ | _____ × | _____ | = _____ |
|  |  |  | _____ Btu/°F |
|  |  |  | TOTAL |

## C. Heat Capacity of Mass Surfaces Enclosing Spaces Connected to Direct Gain Spaces

| MASS DESCRIPTION (INCLUDE THICKNESS) | AREA | UNIT HEAT CAPACITY (TABLE H) | TOTAL HEAT CAPACITY |
|---|---|---|---|
| Trombe Walls _____ | _____ × | 3.8 | = _____ |
| Water Walls _____ | _____ × | 4.2 | = _____ |
| _____ | _____ × | _____ | = _____ |
| _____ | _____ × | _____ | = _____ |
|  |  |  | _____ Btu/°F |
|  |  |  | TOTAL |

## D. Total Heat Capacity

_____ Btu/°F

(A+B+C)Btu/°F

## E. Total Heat Capacity per Square Foot

_____ ÷ _____ = _____ Btu/°F–sf

TOTAL HEAT CAPACITY        CONDITIONED FLOOR AREA

## F. Clear Winter Day Temperature Swing

|  | TOTAL PROJECTED AREA (WORKSHEET II) | COMFORT FACTOR (TABLE I) |  |
|---|---|---|---|
| Direct Gain_____ | _____ × | _____ | = _____ |
| Sunspaces or _____ | _____ × | _____ | = _____ |
| Vented Trombe Walls |  | _____ ÷ _____ = _____ °F |  |
|  |  | TOTAL        TOTAL HEAT CAPACITY |  |

## G. Recommended Maximum Temperature Swing

_____ °F

**Compare Line F to Line G**

FIGURE 7-3

*This worksheet permits the designer to estimate temperature swings
in a house without the back-up heating system operating.*

# Worksheet IV: Summer Cooling Performance Level

## A. Opaque Surfaces

| DESCRIPTION | HEAT LOSS (WORKSHEET I) | | RADIANT BARRIER FACTOR (TABLE J) | | ABSORPTANCE (TABLE K) | | HEAT GAIN FACTOR (TABLE L) | | LOAD |
|---|---|---|---|---|---|---|---|---|---|
| Ceilings/roofs_____ | _____ | × | _____ | × | _____ | × | _____ | = | _____ |
| _____ | _____ | × | _____ | × | _____ | × | _____ | = | _____ |
| _____ | _____ | × | _____ | × | _____ | × | _____ | = | _____ |
| Walls _____ | _____ | × | na | | _____ | × | _____ | = | _____ |
| _____ | _____ | × | na | | _____ | × | _____ | = | _____ |
| Doors_____ | _____ | × | na | | _____ | × | _____ | = | _____ |

_____ kBtu/yr
TOTAL

## B. Non-solar Glazing

| DESCRIPTION | ROUGH FRAME AREA | | NET AREA FACTOR | SHADE FACTOR (TABLE M) | | HEAT GAIN FACTOR (TABLE L) | | LOAD |
|---|---|---|---|---|---|---|---|---|
| North Glass _____ | _____ | × | 0.80 | _____ | × | _____ | = | _____ |
| _____ | _____ | × | 0.80 | _____ | × | _____ | = | _____ |
| East Glass_____ | _____ | × | 0.80 | _____ | × | _____ | = | _____ |
| _____ | _____ | × | 0.80 | _____ | × | _____ | = | _____ |
| West Glass _____ | _____ | × | 0.80 | _____ | × | _____ | = | _____ |
| _____ | _____ | × | 0.80 | _____ | × | _____ | = | _____ |
| Skylights _____ | _____ | × | 0.80 | _____ | × | _____ | = | _____ |
| _____ | _____ | × | 0.80 | _____ | × | _____ | = | _____ |

_____ kBtu/yr
TOTAL

## C. Solar Glazing

| SOLAR SYSTEM DESCRIPTION | ROUGH FRAME AREA | | NET AREA FACTOR | SHADE FACTOR (TABLE M) | | HEAT GAIN FACTOR (TABLE L) | | LOAD |
|---|---|---|---|---|---|---|---|---|
| Direct Gain_____ | _____ | × | 0.80 | _____ | × | _____ | = | _____ |
| _____ | _____ | × | 0.80 | _____ | × | _____ | = | _____ |
| Storage Walls_____ | _____ | × | 0.80 | _____ | × | _____ | = | _____ |
| _____ | _____ | × | 0.80 | _____ | × | _____ | = | _____ |
| Sunspace _____ | _____ | × | 0.80 | _____ | × | _____ | = | _____ |
| _____ | _____ | × | 0.80 | _____ | × | _____ | = | _____ |

_____ kBtu/yr
TOTAL

**D. Internal Gain** _____ + ( _____ × _____ ) = _____ kBtu/yr

CONSTANT COMPONENT (TABLE N)    VARIABLE COMPONENT (TABLE N)    NUMBER OF BEDROOMS

**E. Cooling Lad per Square Foot**    1,000   ×   _____ ÷ _____ = _____ kBtu/yr

(A+B+C+D)    FLOOR AREA

**F. Adjustment for Thermal Mass and Ventilation**    _____ Btu/yr–sf

(TABLE O)

**G. Cooling Performance Level**    _____ Btu/yr–sf

(E-F)

**H. Comparison Cooling Performance** (from previous calculation or from Table P)    _____ Btu/yr–sf

**Compare Line G to Line H**

FIGURE 7-4

*This worksheet is used to calculate natural heating (cooling performance level).*

Temperature swing is a function of several factors, among them passive solar gain and thermal mass. This worksheet asks the designer to provide information on both the incidental and intentional thermal mass in the house design as explained in chapter 3. If not satisfied with the results of this analysis, the designer can achieve a smaller temperature swing by adding more mass. As noted in chapters 1 and 3, additional mass creates a more thermally stable and comfortable interior.

The fourth worksheet, shown in figure 7-4, allows the designer to determine the cooling performance level—that is, the cooling load—and takes into account both internal and external heat gain.

These worksheets duplicate the process of hand calculation described in the previous section. However, *Passive Solar Design Strategies* contains all of the region-specific data a designer needs to complete the calculations. For example, it includes data on R-values of windows and roofs and heating degree day data. The result: You won't need to spend hours poking through reference books or searching through libraries to find the technical information you need to analyze the energy performance of a home design.

*BuilderGuide: The Software Version.* Like the hand-calculation method, the worksheets require a fair amount of time and a considerable number of calculations. For this reason, the Passive Solar Industries Council, now known as the Sustainable Building Industry Council, offers *BuilderGuide*, the software version of the worksheets. Two versions are available, DOS and Windows. (It is not available for Macintosh.)

The DOS program, costing $50, is a computerized spread sheet that walks the designer through the worksheets. The Windows version, costing $100, is a more sophisticated energy analysis program. It allows users to enter data more quickly and with less effort than the worksheets or the DOS version. It is a joy to work with! After the program is loaded, you simply open a new project, select the location, and name the project. The program opens with a screen consisting of a box with eight tabs (figure 7-5). Seven of the tabs open up forms (cards) used to record details of various parts of the building, such as walls, windows, and thermal mass. The eighth permits the user to access the energy-performance program, which determines heating and cooling loads, percent solar heat, back-up heating requirements, and cooling loads.

When entering data on various components of a building, clicking on the blank spaces on the on-screen forms opens a box that displays a number of options. For example, when filling in details on the walls a box appears displaying numerous wall construction options listed along with their R-values. After selecting the type of construction used in the walls, the designer enters the area of each wall or its dimensions and the software calculates its square footage. If the tables don't provide the data you need, you can click on the Construction Library, a bank of information on building components, and add

FIGURE 7-5

BuilderGuide for Windows *opens with a simple screen with eight tabs. Click on each of the first seven, enter the data, and when done, click on the last tab to determine the performance. Values can be changed to enhance year-round performance.*

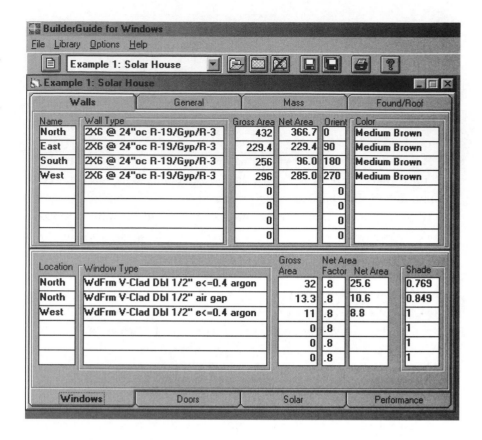

the data to the permanent file. I added information on R-values of straw bale walls and adobe walls, for example, in a few seconds.

Filling out the forms in *BuilderGuide for Windows* requires much less time and effort than filling out the worksheets or completing the DOS spreadsheet program, because the Windows program is highly automated. Once data on walls, windows, mass, and so on has been entered, it is time to run a performance analysis.

BuilderGuide offers two analyses. The first yields the annual heating load (conservation performance level), solar heating potential (as a percentage of total heating load), and back-up heating requirement (auxiliary heat performance level). It also provides the temperature swing and cooling load (summer cooling performance level). That is, the software provides all of the information obtained by filling out the worksheets. Like the worksheets, *BuilderGuide for Windows* provides energy-performance data on the proposed design as well as data on a base case home designed and built using standard construction practices for comparison.

The second analysis provides heat loss through various components: walls, doors, windows, foundation, floor, roof, and via infiltration. This is vital for pinpointing areas in need of improvement.

*BuilderGuide*'s performance analyses, which would take many hours of laborious work by hand, are performed quickly and efficiently. Results can be printed out and saved to disk for future reference. If the results fall short of intended goals, a designer can manipulate various building components to improve the building's long-term energy performance. For example, a designer may increase insulation, shift windows to the south side for enhanced solar gain, and add thermal mass. After these changes are made, another performance analysis can be run and compared with the first attempt. If the proposed changes failed to meet the designers energy goals, additional modifications can be made.

***The Pros and Cons of BuilderGuide.*** *BuilderGuide for Windows* is a valuable design tool, worth its weight in gold. Because the Sustainable Buildings Industry Council provides data on your location, the program saves considerable time and effort.

For an additional cost, users can order information on solar gain and heating degree days for any one of 239 locations in the United States, including numerous sites in Alaska. If your city or town is not in their databank, however, don't despair. You will most likely be able to order software with data on a nearby city or town.

The manual for *BuilderGuide for Windows* is well written. In addition to instructions on running the program, it contains useful examples, background information on various technical terms, and a glossary. The software is user-friendly and powerful. Additional software (WinGuide) can be purchased from SBIC to compare your design to the 1993 Model Energy Code requirements for your area.

As good as *BuilderGuide* is, it does have a few minor "flaws." The manual, for example, has a few minor typographical errors, but nothing that obstructs one's understanding of the material. The sample screen displays in the manual are difficult to read. For example, the fine print on the graphs of performance is too small to read. Another problem is that the print-out of the energy analysis is not the same as the on-screen version. The on-screen version is much more complete.

In addition, *BuilderGuide* does not provide as much data on wall, roof, and floor options as is needed by many designers, and there is no information on alternative building systems such as straw bale or adobe. If you are planning to build a straw bale home or are using components such as structural insulated panels for exterior walls, you will need to supply some of your own information (for example, R-values of walls).

I found the number of framing and roofing options listed in *Passive Solar Design Strategies* to be limited, restricted to the most common building materials and techniques. Data on windows and foundations is also limited. If you are familiar with home construction and have a good grasp of the materials, techniques, and R-values, however, this should pose little trouble. You can add the information on R-values to the construction library. If you are not a veteran designer or builder, you may have to do some research to find the information.

On balance, the problems with *BuilderGuide for Windows* pale in comparison to its utility and value to architects, builders, and owner-builders.

## Energy-10

An even more powerful design tool than *BuilderGuide for Windows* is *Energy-10* software, produced through a partnership of the Sustainable Buildings Industry Council, National Renewable Energy Laboratory, Lawrence Berkeley National Laboratory, and the Berkeley Solar Group under the leadership of NREL's Doug Balcomb. *Energy-10* is designed for use by a broad audience, including architects, engineers, contractors, and builders.

*Energy-10* comes with a fact-filled notebook, *Designing Low-Energy Buildings*. Chapter 1 provides an introduction to passive solar design. Chapter 2 outlines the key principles and practices. Although the information is primarily useful to newcomers, veteran architects and builders may find it to be a good review. I found the coverage far more thorough than that in *Passive Solar Design Strategies*, the book that accompanies the *BuilderGuide* software.

*Designing Low-Energy Buildings* describes the use of *Energy-10* software and contains a lengthy section on sixteen energy-efficient strategies, such as insulation and daylighting, that are used to improve a building's performance. *Energy-10* software comes with an installation manual and other supporting documentation.

*Designing Low-Energy Buildings* and *Energy-10* are "meant to be used in tandem to assist in the design of energy-efficient, low-rise commercial, institutional and residential buildings," according to Sustainable Buildings Industry Council. Although the principles apply to larger buildings, the package is intended for use in designing small and medium-sized buildings, under approximately 10,000 square feet with one or two thermal zones. A thermal zone is a zone heated or cooled to the same level. A two-thermal-zone building, for example, would be an office with an attached warehouse. Note, too, that although the written material focuses primarily on commercial buildings, the information is relevant to residential structures. According to the SBIC, at this writing, of the nearly 2,100 registered users, most of them use the software to design homes.

*Designing Low-Energy Buildings* and *Energy-10* will require quite a few hours of study and use to fully appreciate its full value and to master its intricacies. Those who purchase a copy of the package and register with the Sustainable Buildings Industry Council receive free technical support, regular program updates, notices of new releases, and information on Energy-10 seminars and workshops. Software upgrades are provided for a nominal fee.

**Energy-10:** *A Closer Look. Energy-10* facilitates integrated design and was developed in conjunction with architects, engineers, builders, and utility officials. It permits designers to analyze the effects of numerous design decisions on a building's energy consumption—notably, heating load and cooling load. Like *BuilderGuide for Windows, Energy-10* allows for region-specific design. Key data on temperature, solar radiation, and utility rates is provided on the CD for 239 cities and towns (parent sites) in the United States. Unlike *BuilderGuide,* for which you have to purchase an additional location file for each site, *Energy-10* CD comes with weather data (files) for all 239 sites. In addition, *Energy-10* also comes with a WeatherMaker program and database that allows the user to customize weather information—that is, to use data for the specific site he or she is working on or, at least, much closer to the site than one of the 239 main locations. WeatherMaker contains temperature data for an additional 3,958 locations.

**DESIGN DEVELOPMENT**

**Predesign**—articulate project goals and targets

**Preliminary design**—design decisions are made

**Design development**—design decisions are translated into blueprints

If WeatherMaker doesn't have a file for your precise location or one nearby, the program enables you to generate your own files, using temperature data from a local source, for example, a local weather station. Users in other countries can also create weather files for their sites, making *Energy-10* useful in other parts of the world (much more so than other programs).

Like *BuilderGuide, Energy-10* allows designers to perform energy analyses during the earliest stages of design, the predesign stage, a time in which fundamental project goals and targets are articulated. In a sense, then, these programs allow a user to define and evaluate a new project *before* it has been designed. Far better to design with energy in mind early on, note the authors of *Designing Low-Energy Buildings,* than to perform a comprehensive energy analysis on a completed plan with only intuited or crude energy savings estimates, only to find out that the completed design performs poorly.

*Energy-10* and the accompanying printed material are also useful during the next phase, the preliminary design stage, when basic design decisions are established. Next comes design development, a phase in which the architects translate design ideas into blueprints. All decisions can be checked by *Energy-10* to keep the project on target. *Energy-10* can also be used to analyze decisions made during the construction of a project—for example, if materials or building systems change for some reason.

*How Does* **Energy-10** *Work?* As just noted, *Energy-10* can be used at any stage of design. My description assumes that the software and supporting information are used from the very beginning (predesign) to the very end (construction) of the process.

To begin, the user enters the program and clicks on the new project button. A screen appears that asks for five basic details of the anticipated building (figure 7-6). Let's assume that we're building a single-family residence, a single-

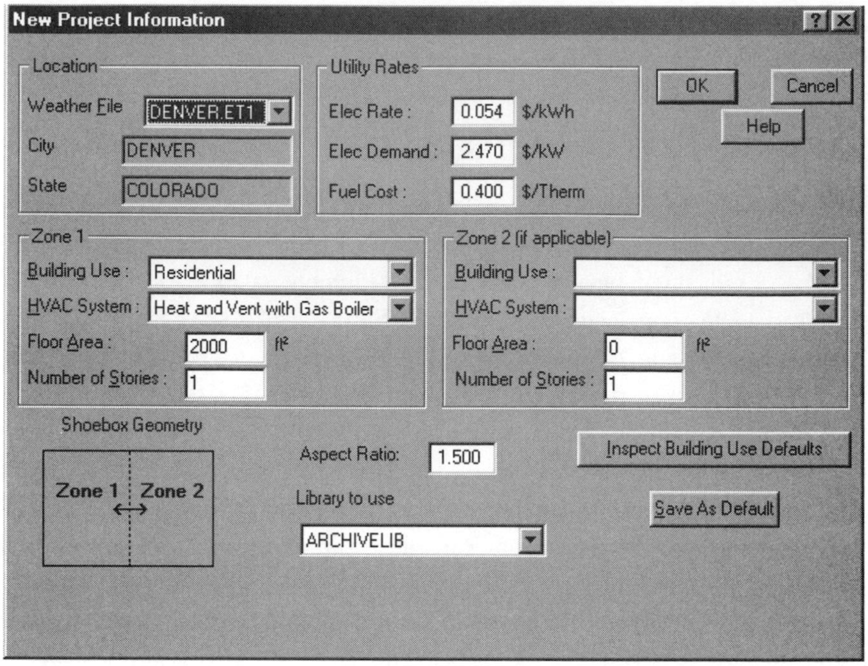

FIGURE 7-6
Energy-10 *commences with minimal data input, as shown here. The computer quickly assembles two buildings, a reference case and a low-energy building.*

zone building. The user selects the appropriate location file, which contains weather data for the building site, then selects the building use, such as residence, school, or office. The user then selects one of the ten defined heating, ventilation, and air-conditioning back-up systems and enters the square footage and the number of stories. (If there is a second zone, similar information is entered for that as well.) Finally, the user clicks on local utility rates and confirms that the listed rates are accurate.

With this input, *Energy-10* quickly assembles a Predesign Reference Building (PRB) using an extensive set of default values. That is, it uses typical construction details and average values for many building characteristics, such as insulation, windows, thermostat set points, and so on.

The Reference Building that *Energy-10* assembles is rectangular with an east-west axis, a shoebox design that is most likely a far cry from what the designer has in mind. The program calculates anticipated heating and cooling loads and delineates how energy use and energy cost are divided among heating, cooling, lights, fans, and plug loads (basically anything that is plugged into an outlet such as appliances and electronic devices). The results can be quite enlightening.

Why not begin with a structure that more closely resembles the final structure? In *Designing Low-Energy Buildings*, the SBIC answers the question this way: In the incipient stages of design, it generally does not make sense to input elaborate building geometries. The Reference Building simulation results provide useful insight quickly and effortlessly.

*Energy-10* simultaneously assembles a second building, derived from the Reference Building. It, too, is a simple shoebox, but is designed with numerous energy-efficiency strategies selected by the user such as daylighting, south-facing glass, and thermal mass. The result is called the Low-Energy Building (LEB).

*Energy-10* performs numerous analyses on both buildings and generates stunning graphs of a variety of energy parameters, including annual energy use for heating, cooling, lighting, and plug load (figure 7-7). These graphs allow

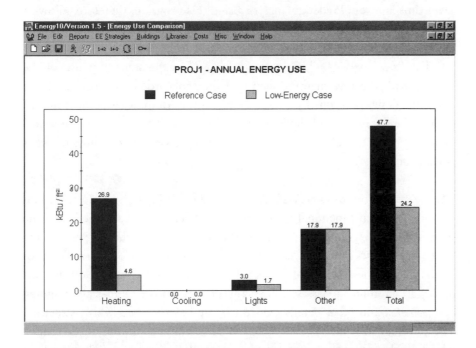

FIGURE 7-7
Energy-10 *permits graphical analysis of many different variables, among them the annual energy use of a building.*

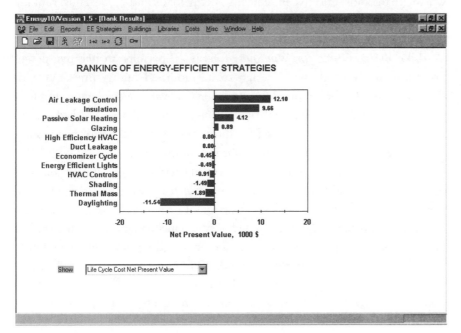

FIGURE 7-8
*Ranking the energy performance of a proposed building helps a designer see where the biggest savings in energy can be made.*

the designer to compare the Reference and Low-Energy Buildings, labeled as Building 1 and Building 2. The side-by-side comparison shows the potential for energy savings available to a designer who goes beyond standard construction practices. In addition, *Energy-10* contains a ranking function that orders the energy savings of various options, showing the most to the least effective, not just for Btus but also with respect to their economic impact (figure 7-8). Ranking performed at the beginning of each project provides guidance and saves time later on. Predesign analysis using this program, therefore, allows a designer to proceed from the predesign to the preliminary design phase knowing which energy strategies will result in the greatest savings in a building.

The Low-Energy Building is typically used as the starting point for the design. With the low-energy design and the rank-ordered energy results in hand, the designer begins to change details of the computer-generated model, such as the dimensions of the building and distribution of windows in accordance with the actual site, owner's requirements, view, and a host of other considerations.

*The Pros and Cons of* **Energy-10.** As the Sustainable Buildings Industry Council notes, "Just as functional or structural considerations are carried from the beginning to the end of a project, energy considerations should pervade the entire design process." The usefulness of *Energy-10* and other similar programs during these stages of project development and construction lies primarily in its ability to assess the energy performance of materials, building components, and other details individually and as part of a whole structure. The program also allows a designer to rank various options to determine which are the most effective in reducing energy loads and costs. Thus, ranking functions help a designer focus on those elements of building design that yield the greatest energy dividends.

Like *BuilderGuide for Windows, Energy-10* permits an iterative design process—that is, it allows a design to be modified many times. As with *BuilderGuide*, this program allows designers to evaluate the energy implications of each and every change they make along the path to a finished design. Each design analysis can be saved for future reference and performance summaries of each design can be printed out for comparison.

*Energy-10* is user friendly and powerful, performing complex analyses quickly in the time it takes to drink a cup of tea. *Energy-10* is well documented, too. The first-time-user instructions in the notebook are extremely helpful. The Sustainable Buildings Industry Council offers a succinct on-line introduction to the program, as well as a more lengthy printed introduction to the software. They also offer on-line slide shows on *Energy-10* and Weather-Maker that provide a useful overview. For those who want formal training, the SBIC offers two-day workshops throughout the United States, often in conjunction with national meetings such as the American Solar Energy Society's

*The Sustainable Buildings Industry Council notes, "Just as functional or structural considerations are carried from the beginning to the end of a project, energy considerations should pervade the entire design process."*

annual meeting or meetings of the American Institute of Architects. *Energy-10* is also taught in more than eighty U.S. colleges and universities.

*Energy-10* provides numerous colorful graphics that display the results of the many analyses it performs. These graphics assist the user in fully understanding the energy implications of his or her design decisions. For professional designers, print-outs of the graphs may prove useful in illustrating the benefits of various energy-efficient design features to clients. Graphs can be easily cut and pasted into word-processing documents for reports to clients or lenders.

*Energy-10* also contains an AutoSize function that sizes heating, air-conditioning, and ventilation systems based on the final design parameters, local climate, and solar potential.

Furthermore, *Energy-10* location files can be customized by Weather-Maker. This makes it useful not only in the United States, but in other countries as well.

*Energy-10* receives high grades from users. It is judged as more user-friendly and more successful than other software by 60 percent of the users, according to surveys by the Sustainable Buildings Industry Council.

*Energy-10* does have a few weaknesses. I found the building options somewhat limited. Roof choices were especially proscribed. Architects and builders of straw bale and other natural homes will need to customize the Construction Library, including data on R-values of their walls, but that's to be expected since these building methods have not yet entered the mainstream.

Another thing to bear in mind is that *Energy-10* provides estimates, not necessarily accurate predictions of future energy use. The reasons for this are many. For one, people work or live differently in a building. Differences in temperature settings and use of windows and lights, for instance, can have a profound effect on how much energy is consumed in a building.

In *Designing Low-Energy Buildings,* the Sustainable Buildings Industry Council notes that "occupant effects" can result in an annual energy use ranging from "70 percent to 140 percent of the average use in commercial and residential buildings." That is, a building can use 30 percent less to 40 percent more energy than predicted, depending on the use patterns. "Some building owners," they note, "keep the thermostat low in the winter and are careful about not leaving doors and windows open in cold weather. They might be the same people who use air-conditioning only when necessary and rely on open windows on mild days in the spring, fall, or summer to keep the building cool by natural ventilation. Others rely on their furnace or air conditioner, paying a higher utility bill. Also, different individuals have different personal comfort zones."

Another reason for potential discrepancies between *Energy-10*'s analyses and the actual energy use of a building is that the software relies on average weather and solar data for its energy estimates. However, as the SBIC points out, "weather conditions during any one year can be much different from the long-term average." With global temperatures rising as a result of human-induced climate change, summer cooling loads may be increasing in many parts of the world while winter heating loads may be declining.

Another factor is internal heat gain. "Internal gains are notoriously unpredictable, especially plug loads. This poses a significant uncertainty to forecasting energy use, more so in commercial buildings than in residences," notes the SBIC.

Yet another reason why projected and actual energy use may not correspond is errors in input data—that is, differences between the description of a building and how it was actually constructed. Furthermore, if insulation isn't installed correctly, the building won't perform as anticipated.

The Sustainable Buildings Industry Council advises professional designers to be careful how they convey *Energy-10* results to clients. For owner-builders, be careful not to expect an exact match between estimated and actual energy consumption and savings. These are estimates based on reliable, conservative, and widely accepted procedures, but they're still estimates.

Another potential problem for some prospective users of *Energy-10* is the software's cost. At this writing (June 2002), the package costs $250 for professionals (nonmembers of SBIC) and $100 for students. Although that may seem like a lot to spend, the package is well worth the money. It could save tens of thousands of dollars over a building's life span. (Note: Readers of this book are entitled to half off. See the coupon at the back of this book.)

*Energy-10* is being revised to include additional features, for example, the use of photovoltaics to produce electricity. Work is underway on graphical sketch capabilities that will allow an architect to alter the design of a two-zone building, for example, to include additional zones. Graphical inputs will then be automatically converted to building descriptions. To learn about new versions, contact the SBIC, listed in the Resource Guide.

At this point, you may be wondering which of the Sustainable Buildings Industry Council's two software programs would be best for you or your designer. When asked this question, then SBIC Associate Director Doug Hargrave responded, "In general, we are trying to steer people to *Energy-10* because the feedback has been that it does everything *BuilderGuide for Windows* does, and it does it more powerfully." Actually, it outperforms *BuilderGuide*, offering more data, more detailed analyses, and a wide range of graphs and comparisons.

### *Solar-5:* A DOS-Based Analysis and Design Tool

Another potentially valuable design and analysis tool is *Solar-5*, a computer program that designers can download free of charge from the Internet by logging on to the University of California in Los Angeles' Architecture and Urban Design web site. *Solar-5* was developed by the UCLA Architectural Design Tool Development Project headed by Professor Murray Milne. Numerous

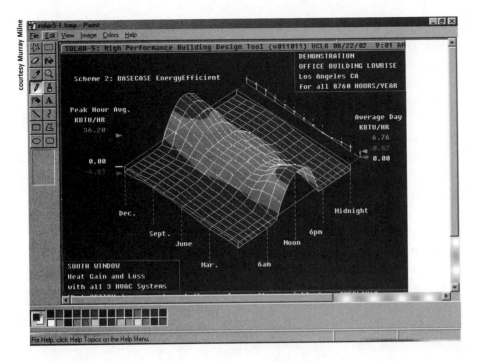

courtesy Murray Milne

**FIGURE 7-9**
*Solar-5 creates stunning graphics of building performance, as shown here.*

## HOME ENERGY EFFICIENT DESIGN (HEED)

The Windows version of *Solar-5,* produced by Professor Milne and his students, is a powerful new tool that surpasses *Solar-5* in many respects. This program can be used to analyze the energy performance of new homes and additions—not only for heating and cooling but also electrical energy for lighting and running appliances. *HEED* also allows individuals to analyze energy savings from energy-efficiency measures, such as window awnings or added insulation, in existing homes and apartments. *HEED* is more user-friendly than *Solar-5,* but the software is currently (as of September 2002) designed for use only in Southern California. Don't be dismayed, however. The creators plan to adapt if for California-wide use next, then to provide data and programming to make it suitable for use throughout the United States. You can download a copy for free at www.aud.ucla.edu/heed.

graduate students have helped to develop the program over the past twenty-five years as their thesis work.

*Solar-5* is a DOS-based program, a bit of an anachronism in this age of point and click, but it is quite powerful; *Solar-5* helps students, architects, builders, and others to understand complex phenomena and offers stunning graphical displays. (As this book goes to press, we learned that a Windows version of *Solar-5,* called *Home Energy Efficient Design,* has been released. See sidebar.)

Like *Energy-10,* the user begins by entering basic data on the building, including the location, building type, floor space, number of stories, and project name. The program then designs a base case, a "good" building in your climate, or a superefficient building, a high-performance, energy-efficient building—and it does so very quickly using defaulted values. *Solar-5* then provides numerous graphs that illustrate the building's energy performance (figure 7-9). The designer manipulates the design in accordance with project requirements.

*Solar-5* uses average monthly data for 239 locations. If your site isn't represented in its database, you can enter your own data. Like *Energy-10, Solar-5* analyzes the performance of the base case building, or any modifications you make, by running through every hour of every day of a full year. Also like *Energy-10, Solar-5* contains a construction library that lists numerous building materials and components that can be included in the design to enhance energy performance. Virtually all passive design strategies such as direct gain and thermal storage walls can be simulated with this program. The construction library contains many different kinds of windows and many thermal mass options as well. You can add to the library, if necessary.

*Solar-5* automatically keeps track of all the designs and modifications made. The program even allows the designer to add awnings to reduce solar gain in the summer and transitional seasons and takes this into account when running its energy analysis—a feature not included in *Energy-10* or *Builder-Guide for Windows*. *Solar-5* also takes into account seasonal differences in the yard that affect a building's performance, for example, snow cover during the winter and green grass in the summer, not available in the other two programs.

*Solar-5*, like *Energy-10*, pinpoints the most cost-effective ways of saving energy in a new design. It will, for instance, show precisely where heat gain and heat loss are expected to occur during the hottest and coldest parts of the year. Working with this information, the designer can make changes to improve year-round performance.

Complex floor plans are accommodated by *Solar-5*, as are numerous thermal zones, giving it an advantage over previously discussed software. *Solar-5* also allows the designer to compare the performance of up to nine designs by 48 different criteria, such as annual heat load. In addition, this software permits the study of the interaction of various elements, for example, weather and air infiltration. The newest version permits the user to determine how automatic light controls (e.g., autodimming) can trim energy use and reduce greenhouse gas emissions.

***The Pros and Cons of* Solar-5.** Professor Milne characterizes *Solar-5* as "fast and user-friendly." I downloaded the program in a few minutes and found that it was fast and was relatively easy to learn. However, in my view *Solar-5* is much less user-friendly and more difficult to master than either *BuilderGuide for Windows* or *Energy-10*. It comes with an electronic tutorial, but the tutorial is more a summary of the graphical interpretation than instruction on the use of the program. I would have preferred a tutorial that walked me through the design and analysis of a house. *Solar-5* also comes with a user's manual, which the user can download and print for later use. It is a helpful reference guide.

Professor Milne also provides other software for designers, including *Climate Consultant*, which graphs climate data in various locations in dozens of ways that many will find useful. This program performs an analysis that recommends the most appropriate passive design strategy for a particular area.

*Solar-2*, another computer program from UCLA, plots sunlight penetrating through a window with various combinations of overhangs and rectangular fins to control daylighting. It also plots an hour-by-hour "sun's eye" view of a building. It will print tables that show the percentage of windows in full sun and solar radiation on glass, among other useful information.

Yet another computer program available from UCLA is *Opaque*. This software draws a detail of wall or roof sections, then calculates U-values. It plots temperature difference across the wall, heat flow through the envelope, and more. Like the other software available from UCLA, it can be downloaded free of charge from the web site.

## *EnergyPlus:* DOE's Most Recent Software Tool

Yet another potentially useful tool is *EnergyPlus*, released in April 2001 by the U.S. Department of Energy. This program is designed for building professionals: architects, builders, engineers, and building owners and managers. Like *Energy-10*, it takes a whole-building approach that is useful during the planning, design, and construction of a home. Also, like *Energy-10*, the program is designed to help cut energy use in buildings. It contains a simulation code that allows a user to predict the impacts of many factors, such as windows and ventilation equipment, on energy efficiency and occupant comfort. Users can even simulate the effects of window blinds, special electrochromic window glazing, and daylighting systems on energy performance.

*EnergyPlus* is based on two previous programs, *BLAST* and *DOE-2*, released in the early 1980s. Although these programs are considered by many to be the most comprehensive and accurate energy simulation programs available, they are not very user-friendly, although dedicated programmers voluntarily produced kinder, gentler versions with good graphical interfaces (version *Power-DOE*, for instance). In addition to this weakness, *BLAST* more accurately simulated thermal mass, while *DOE-2* more accurately simulated HVAC systems. Both were limited in graphic representation of results.

Extensive revisions of these programs, involving a massive rewrite of the code, resulted in *EnergyPlus*, a program with even greater simulation capacity than its predecessors combined. Today, *EnergyPlus* is one of the most sophisticated and comprehensive tools on the market. New capabilities are planned, including simulations that illustrate the costs and benefits of solar thermal (solar hot-water systems) and solar electricity. In its April 2001 version, however, *EnergyPlus* reads and writes from text files—that is, it looks more like a DOS-based program than a Windows-based program, and suffers from lack of a user-friendly interface. Dedicated users will no doubt tackle that problem.

*EnergyPlus* and its user manual are available free of charge on-line, although at typical modem speeds downloading the software can take over an hour. Periodic updates are posted on the Internet at the DOE site and are published in the *Building Energy Simulation User News*.

*EnergyPlus* is flexible and powerful. It provides users an opportunity to answer very specific design questions. It is accurate and detailed. Although it is primarily a simulation engine, it does link to other programs, a feature that extends its capabilities and power. This program focuses on design issues and their economic and environmental impacts, as well as their effects on the comfort of occupants.

I found *EnergyPlus* much more difficult to master than *Energy-10*, and less user-friendly, but to architects savvy in computer simulation software its power and sophistication make it well worth the effort.

## DESIGN TOOLS FOR BETTER BUILDINGS

Years of experience with passive solar heating and cooling have taught many architects and builders the keys to successful passive design. But the learning process has been a costly one: a legacy of passive solar homes that provided fewer benefits than promised. Occupants of these homes suffered unnecessary discomfort and higher-than-desirable utility bills. These problems stem, in large part, from the fact that many of these homes were designed and built in an era lacking important tools of integrated design that permit a designer to determine requirements for mass, insulation, glazing, and other elements.

Fortunately, the solar design landscape has changed by the introduction of energy-analysis tools. Whether by hand calculation, worksheets, or sophisticated software, energy analysis is helping designers improve year-round performance and comfort of their buildings. As we have seen, energy-analysis software is the fastest and most powerful tool on the market. It not only helps designers engage in integrated design, it can assist in holding down construction costs by focusing attention on components of a building that offer the greatest energy savings.

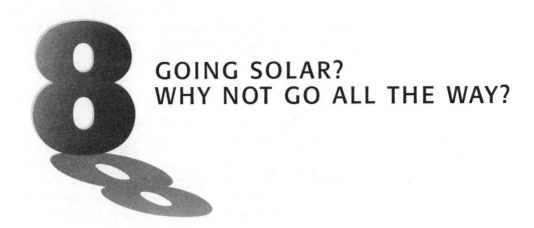

# GOING SOLAR?
# WHY NOT GO ALL THE WAY?

SINCE THE INFAMOUS oil embargo of the early 1970s, the promise of renewable energy technologies has waxed and waned in a disheartening fashion. In the 1990s, however, renewable energy experienced a welcome upswing. In fact, the news on several renewable energy technologies was nothing short of remarkable. Wind energy, for example, became the world's fastest-growing source of electricity, gusting from a paltry 10 megawatts of total generating capacity in 1980 to nearly 15,600 megawatts in 1999, or the equivalent of fifteen large nuclear power plants. In 2000, wind power grew again by an astounding 6,500 megawatts!

Second place in the energy race did not go to coal or natural gas or nuclear power, but to photovoltaics (PV)—solar electricity. According to the Worldwatch Institute in Washington, D.C., PV production in the 1990s grew at a rate of 16 percent per year! British Petroleum, the third largest oil company in the world, has ventured into renewable energy by purchasing one of the nation's foremost manufacturers of photovoltaics, Solarex. While still far behind wind energy in total output, the growth in photovoltaic sales is impressive and promises to continue well into the new millennium.

Passive solar design and the implementation of energy-efficiency measures in homes and businesses also experienced a dramatic upsurge of interest and enthusiasm. Numerous colleges, universities, and corporations built new facilities that meet substantial portions of their heating and cooling loads passively.

Slowly, and clearly, the energy picture is changing. Increasing interest in passive solar heating and cooling and the resurgence in electricity generation from renewable sources, combined with declining oil and natural gas supplies, growing evidence that the combustion of fossil fuels is severely altering the planet's climate, and the ever-present threat of terrorism could signal the dawn of a new era of clean energy. The shift to clean, renewable energy would help address a wide assortment of environmental ills, including urban air pollution, acid deposition, global climate change, oil spills, and habitat loss, which result from the production, processing, transportation, and consumption of fossil fuels.

But are these changes enough? Will the solar revolution we're witnessing be sufficient to create a sustainable future?

Not by a long shot.

## CREATING SUSTAINABLE SHELTER

Although solar and wind energy are crucial components of a sustainable world, renewable energy is only one factor in a complex equation of sustainability. Nowhere is this more evident than in home construction.

Each year, 1.2 million new homes are built in the United States to accommodate our growing population, according to the National Association of Home Builders. New single-family homes, averaging about 2,200 square feet apiece, each require 14,300 board feet of framing lumber. If laid end to end, the 2 × 4s and 2 × 10s used for framing walls, floors, and roofs in one house would extend about two and a half miles. If the lumber for all of the new homes built in America were laid end to end the line would stretch 3.2 million miles—to the moon and back six and half times. Sheathing materials such as plywood and oriented strand board for a single-family dwelling would cover an area twice the size of a tennis court, or about 6,300 square feet. Multiplying that by 1.2 million homes each year helps us see the enormous quantity of resources required to satisfy our ever-growing need for new housing—and these are just two of the many products that are used to build most of the nation's new homes.

Conventionally built, passively heated, and passively cooled homes, while efficient and primarily reliant on renewable energy, also contribute to the drain on the planet's resources. Unless they're built with natural materials such as adobe, straw, or sustainably harvested lumber, new homes require massive clear cuts and expansive mines to supply building materials and furnishings. The production of these materials consumes huge amounts of fossil-fuel energy. Moreover, new homes often usurp farmland, open space, and wildlife habitat.

Conventionally built, passively conditioned homes also produce billions of gallons of sewage that are emptied into septic tanks or municipal sewer systems. Septic tank leach fields deposit huge amounts of waste just below the ground's surface, and serve as a potential source of surface and groundwater pollution. Although sewage from municipalities is treated to reduce its waste content, effluents are often laden with pollutants that decrease water quality in surface waters.

Although modern passive homes and the families that live in them are emancipated to varying degrees from utility companies and mechanical heat-

FIGURE 8-1
*Exterior view of author's passive solar/solar electric home, built from many green building materials.*

FIGURE 8-2
*Interior view of the author's home, showing cabinets made from salvaged lumber and tile made from mine waste.*

ing and cooling systems, they still depend heavily on an extensive and costly infrastructure to provide food, electricity, fresh water, and waste disposal. Although they have achieved some independence through passive heating and cooling, most of these homes would cease functioning should the resource supply infrastructure be cut off.

Thankfully, there is a movement afoot to change all of this, to create truly sustainable shelter. The movement has been led by a growing cadre of natural builders, people building homes out of natural materials. It is now being joined by many more conventional homebuilders. They realize that although renewable energy technologies are important, indeed essential, to a sustainable future, they are only part of the equation.

I learned this when I embarked on the construction of my current home (figure 8-1). Dedicated to making the home as energy independent as possible, I designed a passively heated and cooled structure supplied with electricity exclusively from wind and sunlight. We built the house largely out of 800 rammed earth tires, which form the foundation and walls of the structure. We built a living room from straw bales and used numerous recycled, salvaged, or green building materials, including carpeting, carpet pad, and insulation made from recycled materials (figure 8-2). The house recycles some of its graywater and captures rainwater to use for bathing and washing dishes. We recycled almost all building waste and even planted native grasses around the house after construction was completed.

Along the way, I learned a great deal about building sustainable shelter. Here are the main points I share in my speeches and workshops:

## SITE SELECTION

To create sustainable shelter, the first step is to select an appropriate site. Many advocates of environmentally sustainable building recommend finding a lot within existing city limits to protect open space, farmland, and wildlife habitat. As David Pearson points out in his book, *The New Natural House*, "It is much more ecologically sensitive . . . to improve the environment of the cities we have and live better in them" than to "lose more countryside to yet more development." More compact settlement patterns are easier and more efficient to service with amenities such as mass transit and fire protection than dispersed development, now commonly known as sprawl. Building a home in a region serviced by mass transit provides an opportunity to leave your car at home, cutting down on mileage, fuel bills, maintenance, and air pollution.

If you must commute by car, living close results in shorter commute times. Reducing time spent behind the wheel of a car, locked in traffic, frees up more time to be with friends and family. And, of course, less driving means less pollution.

If country living is your dream, choose wisely. Select a site that has good wind or solar potential. If possible, purchase a lot that will permit your home to be sensitively nestled in the landscape, out of view and out of ecologically sensitive areas. Leave as much open space as possible. With a little work, you can create your own little nature reserve.

In rural sites, many people make the mistake of locating their home in the most beautiful part of the site. In the process, they destroy the very beauty they sought. A far more sensitive approach would be to leave alone areas that are most precious. Build a house in a part of the site that is least appealing, then turn it into a garden through landscaping. Look out on the beauty you've preserved.

Because site selection is a complex subject, read some books on alternative or natural building to learn more about what you can do to create environmentally sensitive shelter.

## EFFICIENCY

As noted in chapter 2, efficiency is essential to creating truly sustainable shelter, not just energy efficiency, but the efficient use of all resources, such as water and building materials. Build efficiently. If you are framing your home with conventional lumber, studs placed 24 inches on center (rather than the conventional 16 inches on center) will generally provide adequate structural support and will cut lumber use substantially. When designing a home to be built with conventional materials, pay close attention to dimensions to avoid wasted wood. Odd-sized rooms result in more waste than rooms sized to correspond to the dimensions of conventional materials.

A passively conditioned home can be made efficient in the use of water by installing low-flush toilets, composting toilets, low-flow showerheads, and water-efficient washing machines and dishwashers.

## BUILDING SMALL

Creating sustainable shelter also requires serious consideration of building size. Ultimately, we need to rethink our "bigger is better" mentality. In these days of shrinking resources and increasing population, trophy homes are a luxury that come with an enormous environmental price tag—a huge price to pay for extravagance, self-indulgence, and bragging rights. What good is a big, expensive home if you haven't got a decent planet to put it on?

Building large homes not only requires an extraordinary amount of material and energy for construction, it also requires huge amounts of energy for heating and cooling the living space. It requires paints and stains and extensive furnishings, such as carpets and window coverings, the production of which has a destructive impact on the environment. Compact house designs may not show off a person's wealth, but they do illustrate frugality, wisdom, and commitment to an environmentally conscious lifestyle.

With careful design, you can live well in a compact home. Instead of 3,000 to 5,000 square feet, think 1,200 to 2,400 square feet. To cut down on excess square footage, consider scrapping the formal living room—who ever uses it?—and the formal dining room. Although some architects may roll their eyes at these ideas and try to talk you out of them, most of us have little use for formal dining and living rooms. A cozy family room and a family dining area near the kitchen suffice well. At every occasion, I urge designers to find ways to make more efficient use of space. If the idea of living in a compact house appeals to you, there are hundreds of design tricks to get the most out of limited floor space and at least a half-dozen books on the subject to assist you in your task.

## NATURAL BUILDING MATERIALS

Another measure you should consider is building with natural materials—for example, earth or straw, or some combination of the two. Natural homes are clean, comfortable, and beautiful—even breathtaking (figure 8-3). The thick, solid walls create a sense of

## DESIGNING WITH NATURE

In designing and building my home, I tried to abide by the biological principles of sustainability: conservation (use what you need, use it efficiently), recycling, renewable resource use, and restoration. Below are some examples.

### Conservation/Efficiency

Highly insulated walls and ceilings

Insulation around foundation

Earth-sheltered design

Energy-efficient windows

Energy-efficient appliances

Energy-efficient electronic equipment

Compact fluorescent light bulbs

Outdoor clothesline

Insulated slider panels for skylights

Insulated shades for windows

Water-efficient showerhead and toilets

Water-efficient irrigation system for garden

Air locks

Wooden I-beam roof rafters

### Recycling

Packed-tire walls

Recycled carpet, carpet pad, and tile

Recycled cellulose insulation

Salvaged doors

Salvaged wood for cabinets and vanities

Reclaimed paint

Recycled asphalt driveway

Built-in recycling center

Recycled concrete block for planters

Recycled waste from job site

### Renewable Resource Use

Passive solar heating and cooling

PVs and wind generator for electricity

Straw bale walls

Light straw-clay generator shed

### Restoration

Replanted site with native vegetation

*Natural homes are not only good for our physical and mental health, they are exceedingly beneficial to the environment. In taking care of ourselves, we have an opportunity to help safeguard the Earth's ecosystems, the life-support system of our planet.*

comfort and security. Natural homes provide comfort and beauty that often greatly exceed that of conventional modern homes.

All natural building materials are renewable resources. Wood and straw, for instance, are regenerated by natural processes. Even soil is renewable.

Natural homes are also energy efficient and ideally suited for passive solar heating and cooling. Most natural building techniques provide the thermal mass or insulation required to stay cool in the summer and warm in the winter. Because they are typically free of toxic substances found in conventional homes, they are ideal for airtight designs.

Natural homes use less fossil-fuel energy to operate, but also require substantially less energy to build than conventional stick-frame homes popping up by the millions, because they use building materials that are locally harvested and thus have a low embodied energy. Reduced energy consumption provides a wide range of environmental benefits, including reductions in greenhouse gas emissions in a world growing hotter by the year.

Natural building is gaining respectability among professionals, too. More and more architects and builders are emerging who specialize in natural building. Bruce King, a California structural engineer who founded the Ecological Building Network, and others are studying the structural properties of natural building materials in an attempt to better understand and predict their performance.

Natural building techniques are becoming more widely accepted, too, as test results on fire safety and structural strength are made known to code officials. Bank loans and insurance for these buildings are now widely available. And the resale values of these homes can be comparable to conventional homes. As energy prices escalate, their value should increase even more.

FIGURE 8-3
*This charming straw bale home in Rico, Colorado, is heated and cooled passively.*

Readers who want to learn more about natural building can read my book, *The Natural House: A Complete Guide to Healthy, Energy-Efficient, Environmental Homes*. It discusses thirteen natural building techniques, examining how they are built and the pros and cons of each one. It also contains an extensive Resource Guide, which is updated on my Web site: www.chelseagreen.com/Chiras.

## GREEN BUILDING MATERIALS

To create a more environmentally sustainable home, you can select from a growing number of healthy, green building materials. The term "Green building materials" encompasses a wide variety of products, some with low embodied energy, others manufactured from recycled scrap, and others made from natural materials such as cotton or wool. Still others are designed for low toxicity. Many offer two or more of these environmental attributes.

Fortunately, hundreds of green building materials are on the market today, and some are even sold at conventional building supply outlets. Others are available in specialty stores such as Planetary Solutions in Boulder, Colorado, or Eco Wise in Austin, Texas—just two of the nation's nine regional supply outlets.

What is remarkable is that today, every conventional building product has an environmentally benign substitute. From roof shingles manufactured from recycled rubber or plastic to framing lumber from certified forests, to carpeting made from recycled plastic pop bottles, to low-formaldehyde oriented strand board for exterior sheathing, to bamboo flooring to cabinets finished with nontoxic stains and paints, to nails made from recycled scrap, the choices are phenomenal and growing by leaps and bounds. All of these products, and more, are either better for the environment, or gentler on our health.

How does one go about assessing the "greenness" or sustainability of a product? The sidebar on page 244 provides a quick summary of criteria for evaluating products. Making decisions about such products can be complex. Some materials may meet the criteria, but cost too much, either to purchase or to ship to your site. In some instances, a product may meet only one or two sustainability criteria. Nonetheless, that is an improvement over conventional building materials. Some products may require special skills for installation and repair, which are not available in your area. As Sam Clark points out in *The Independent Builder*, it is "not enough for the technology to be possible. Delivery and support systems have to be there too, including . . . local tradespeople who can install and service products."

No one should lead you to believe that choosing environmentally friendly products is going to be simple. However, selecting a suitable material is getting easier every day. Access is improving. Selection is broadening. Information on alternatives is being disseminated by more and more sources.

### NATURAL BUILDING OPTIONS

Straw bale
Straw-clay
Adobe
Cob (monolithic adobe)
Rammed earth
Rammed earth tires
Earthbags
Cast earth
Stone
Cordwood
Log
Papercrete

My advice is to select the products and materials that meet the greatest possible number of the criteria. Don't expect perfection; you won't find it. Virtually all products have some shortcoming. Choose those materials that offer the greatest gain for the environment and the health and welfare of the occupants of the house. One useful strategy is to concentrate on big-ticket items, products that are used in great quantity such as framing lumber, insulation, tile, concrete, and drywall. Purchase as many of these green building products as you can afford.

You may be surprised to find that many green building products are cost competitive. I know I was. In my house, tile, insulation, carpeting, carpet pad, paint, and cabinetry were all comparable in cost to less healthy and environmentally harmful conventional options. If a product does cost more, however, don't dismiss it. The price difference may be offset by savings on another product or by reductions in energy costs. Higher costs may increase comfort levels or reduce health impacts. Price may also reflect quality. Keep track of materials on which you save money, then apply savings to items that cost more.

Don't let builders who are unfamiliar with the wide array of environmentally friendly materials dissuade you. Remember, builders are a conservative group. They build with materials that work for them and others—materials with a tradition of acceptable performance. Builders aren't inclined to experiment on homes. It is too costly if they make a mistake. They do not want home buyers calling them to complain about some product that did not work. They do not want to tear out a new product because it failed to live up to a manufacturer's claims.

Do your own research. Check out the materials. Talk to others who have used them. I've listed buying guides in the Resource Guide at the end of the book.

## ALTERNATIVE WATER SUPPLY SYSTEMS

Another important step on the road to sustainable housing is the installation of alternative means of supplying water for domestic and outside use. One of the most popular is a catchwater system, in which water is captured from roofs and stored in huge tanks or cisterns. When needed, it is pumped into a house, then filtered.

In areas where groundwater supplies are declining and surface waters are drying up due to excessive demand and a lack of rainfall, catchwater is proving to be a great alternative. It not only provides water for our homes, it reduces our impact on groundwater and surface water supplies, reduces energy demand (there is no need to pump water from a well or from a distant water supply), and fosters self-reliance in a world where dependence on centralized production of valuable resources makes us vulnerable to accidents, systemic malfunctions, and the devious actions of others.

Combined with water conservation measures such as efficient shower-heads, low-flush toilets, and superefficient dishwashers and clothes washers, catchwater systems can supply most, if not all, of your needs. To make one work for you, you will need ample rain, a large, relatively clean surface to catch water (steel roofs are ideal), a means to store water (cistern), and a filtration system to remove potential contaminants. Rather than letting water flow off our roofs down the sidewalks and into storm sewers, causing flooding in nearby streams, why not capture the water and use it for cooking, bathing, and watering your garden?

## ALTERNATIVE WASTEWATER SYSTEMS

To help create a sustainable home, many people advocate alternative waste-water systems to reduce or eliminate the depletion and contamination of ground and surface waters. Simple, effective systems have been devised to recycle graywater and blackwater.

Graywater is "waste" water from sinks, washing machines, and showers. Constituting 50 to 80 percent of the water leaving a house in its sewer lines, graywater contains valuable resources, including plant nutrients, that can be put to good use. Recycling graywater also cuts down on demand for water from wells, city mains, or catchwater systems.

The best time to design a graywater system is when you are building a house so you can design one into your home and operate it primarily, or perhaps entirely, by gravity flow.

For a successful system, you'll probably have to change some habits. One of the biggest changes will be replacing that arsenal of toxic cleaning agents under your kitchen sink, which will kill plants, with biocompatible soaps and detergents.

The simplest graywater system consists of a hose from your washing machine to the yard. Although effective, it has its drawbacks. Much more advanced is the system pioneered by Solar Survival Architecture in Taos, New Mexico. They have designed a simple, elegant, and effective indoor planter system to treat graywater. Graywater flows into a lined planter, then courses through a bed of pumice containing billions of microbes that gobble up the waste. Plants in an overlying layer of soil send roots into the pumice that absorb nutrients (the stuff we think of as waste) and water. By the time the graywater reaches the last compartment in this self-contained biological treatment plant, it is relatively clean and can be used to water outdoor plants or to flush toilets, although that requires additional plumbing and can be costly.

Graywater systems vary in complexity, and there are dozens of ways to mess up, so as with my other suggestions, read articles and books, study videos, and consult with experts. Log on to Art Ludwig's Web site (www.oasis-design.net) for a discussion of all the things to avoid.

**FIGURE 8-4**

*Leachate from this solar-heated septic tank drains into a lined planter to the right. Here it flows into a layer of pumice where wastes are broken down. Plants growing in the soil over this layer send roots int the pumice, drinking up the water and nutrients.*

The next step in water recycling is to utilize the nutrients in blackwater, primarily toilet water and water from kitchen sinks. One option is a composting toilet. A well-maintained composting toilet produces little, if any, odor while swiftly converting human waste into a relatively rich, organic matter that can be used to replenish soils. Composting toilets have proven effective in Scandinavia in rural areas where soil conditions are unsuitable for septic tanks. They are also growing in popularity in the United States. Although regulations are pretty tough in the United States, requiring burial of compost or removal by a licensed septage hauler, some people think that properly composted waste can be used in flower gardens or even vegetable gardens. Research this subject carefully, before you put your money down. Joseph Jenkins' book, *The Humanure Handbook,* is a good source of ideas and information.

Blackwater can also be treated in artificial wetlands. These are effective and safe, if properly designed. The most popular is a subsurface system in which sewage flows into a wetland built in a pit partially filled with gravel. Like the pumice in the graywater systems just described, the rocks provide a surface for microbial decay of organic compounds in the waste. Roots of plants growing above the gravel layer draw off the water and nutrients they need to flourish. The naturally purified effluent could be piped to water outdoor plants.

Another blackwater system that is being tried in Colorado and New Mexico by Earthship builders sends waste first to a solar-heated septic tank (figure 8-4). The solids precipitate out and the remaining liquid drains into a sealed planter, much like the graywater planter, but located outside the home. Here,

**FIGURE 8-5**

*The Watson wick filter, an inexpensive replacement for septic tanks.*

the wastewater is purified, reaching levels of cleanliness that would make a sewage treatment plant operator ecstatic.

Tom Watson of New Mexico has devised an ingenious replacement for leach fields and septic tanks, known as the Watson wick filter (figure 8-5). This is an outdoor shallow pumice bed fed by gray- and blackwater. Waste flows first into a small plastic structure, called an *infiltrator,* then seeps laterally into the pumice bed where bacteria and other microorganisms degrade organic matter. Plant roots absorb moisture and nutrients.

## LANDSCAPE FOR ENVIRONMENT AND ENERGY

On any site, care must be taken to minimize erosion and to protect natural vegetation during construction. Be sure to restore the site as well. Using native species is often the best approach, as native species are far easier to establish and keep alive than non-native species. They are already adapted to the local soil and climate. Exotic species of plants may be widely available, but they often require extraordinary amounts of water, fertilizer, and pesticides to maintain. If you are building in a semiarid or arid region, sustainable landscaping dictates the use of xeric plants, plants adapted to very low water conditions. This does not necessarily mean you will be living among cacti and succulents. Many low-water or xeric plants produce luxuriant growth and brilliant flowers that provide nectar for bees and humming birds. The rule here is simple: Work with nature, she is far more forgiving of those who seek harmony—and she knows her business better than the best-trained horticulturist.

Landscaping can also be used to enhance the performance of a home, keeping it cooler in the summer and warmer in the winter, as noted in chapter 5. Properly designed windbreaks, for instance, block cold winds that rob heat from a house, while beautifying a home and providing food and shelter for an assortment of beneficial birds and animals. Shade trees and other types of vegetation cool a home naturally and inexpensively. Carefully selected and planted vegetation combined with thick insulation, overhangs, and earth sheltering can eliminate the need for costly, energy-guzzling air conditioners.

## ONE HOME AT A TIME

Sustainable home building is vital to the future of humankind, and there is a great deal more to learn about it and significant hurdles to overcome. Obtaining permits for designs that include some of the ideas and technologies discussed here, such as graywater and blackwater systems, may be a problem, but you may find that your building department is open to letting you experiment with some of them. Government agencies have even been known to ask to install monitoring equipment to see how the systems are working, so they can assess their performance. If it works out well, authorities are more likely to permit others to follow in your footsteps.

You may also have trouble locating a qualified natural builder and finding green building materials. Prices may be a bit higher than conventional materials. If you build a small, efficient home, though, some of the initially higher expenses required to make a home healthy for people and the planet can be offset by long-term savings.

At times, environmental problems may seem overwhelming and your part in them may seem insignificant. But I encourage you to be neither cynical, skeptical, nor apathetic. You can help create a sustainable future, one home at a time.

# APPENDIX:
# MEAN PERCENTAGE OF POSSIBLE SUNSHINE FOR SELECTED CITIES IN THE U.S. AND CANADA

**BASED ON** period of record through December 1959, except in a few instances. These charts and tabulation are derived from the "Normals, Means, and Extremes" table in U.S. Weather Bureau publication Local Climatological Data.

| STATE/PROVINCE & CITY | JAN | FEB | MAR | APR | MAY | JUNE | JULY | AUG | SEPT | OCT | NOV | DEC |
|---|---|---|---|---|---|---|---|---|---|---|---|---|
| **ALABAMA** | | | | | | | | | | | | |
| Birmingham | 43 | 49 | 56 | 63 | 66 | 67 | 62 | 65 | 66 | 67 | 58 | 44 |
| Montgomery | 51 | 53 | 61 | 69 | 73 | 72 | 66 | 69 | 69 | 71 | 64 | 48 |
| **ALASKA** | | | | | | | | | | | | |
| Anchorage | 39 | 46 | 56 | 58 | 50 | 51 | 45 | 39 | 35 | 32 | 33 | 29 |
| Fairbanks | 34 | 50 | 61 | 68 | 55 | 53 | 45 | 35 | 31 | 28 | 38 | 29 |
| Juneau | 30 | 32 | 39 | 37 | 34 | 35 | 28 | 30 | 25 | 18 | 21 | 18 |
| Nome | 44 | 46 | 48 | 53 | 51 | 48 | 32 | 26 | 34 | 35 | 36 | 30 |
| **ARIZONA** | | | | | | | | | | | | |
| Phoenix | 76 | 79 | 83 | 88 | 93 | 94 | 84 | 84 | 89 | 88 | 84 | 77 |
| Yuma | 83 | 87 | 91 | 94 | 97 | 98 | 92 | 91 | 93 | 93 | 90 | 83 |
| **ARKANSAS** | | | | | | | | | | | | |
| Little Rock | 44 | 53 | 57 | 62 | 67 | 72 | 71 | 73 | 71 | 74 | 58 | 47 |
| **CALIFORNIA** | | | | | | | | | | | | |
| Eureka | 40 | 44 | 50 | 53 | 54 | 56 | 51 | 46 | 52 | 48 | 42 | 39 |
| Fresno | 46 | 63 | 72 | 83 | 89 | 94 | 97 | 97 | 93 | 87 | 73 | 47 |
| Los Angeles | 70 | 69 | 70 | 67 | 68 | 69 | 80 | 81 | 80 | 76 | 79 | 72 |
| Red Bluff | 50 | 60 | 65 | 75 | 79 | 86 | 95 | 94 | 89 | 77 | 64 | 50 |
| Sacramento | 44 | 57 | 67 | 76 | 82 | 90 | 96 | 95 | 92 | 82 | 65 | 44 |
| San Diego | 68 | 67 | 68 | 66 | 60 | 60 | 67 | 70 | 70 | 70 | 76 | 71 |
| San Francisco | 53 | 57 | 63 | 69 | 70 | 75 | 68 | 63 | 70 | 70 | 62 | 54 |

| STATE/PROVINCE & CITY | JAN | FEB | MAR | APR | MAY | JUNE | JULY | AUG | SEPT | OCT | NOV | DEC |
|---|---|---|---|---|---|---|---|---|---|---|---|---|
| **COLORADO** | | | | | | | | | | | | |
| Denver | 67 | 67 | 65 | 63 | 61 | 69 | 68 | 68 | 71 | 71 | 67 | 65 |
| Grand Junction | 58 | 62 | 64 | 67 | 71 | 79 | 76 | 72 | 77 | 74 | 67 | 58 |
| **CONNECTICUT** | | | | | | | | | | | | |
| Hartford | 46 | 55 | 56 | 54 | 57 | 60 | 62 | 60 | 57 | 55 | 46 | 46 |
| **DISTRICT OF COLUMBIA** | | | | | | | | | | | | |
| Washington | 46 | 53 | 56 | 57 | 61 | 64 | 64 | 62 | 62 | 61 | 54 | 47 |
| **FLORIDA** | | | | | | | | | | | | |
| Apalachicola | 59 | 62 | 62 | 71 | 77 | 70 | 64 | 63 | 62 | 74 | 66 | 53 |
| Jacksonville | 58 | 59 | 66 | 71 | 71 | 63 | 62 | 63 | 58 | 58 | 61 | 53 |
| Key West | 68 | 75 | 78 | 78 | 76 | 70 | 69 | 71 | 65 | 65 | 69 | 66 |
| Miami Beach | 66 | 72 | 73 | 73 | 68 | 62 | 65 | 67 | 62 | 62 | 65 | 65 |
| Tampa | 63 | 67 | 71 | 74 | 75 | 66 | 61 | 64 | 64 | 67 | 67 | 61 |
| **GEORGIA** | | | | | | | | | | | | |
| Atlanta | 48 | 53 | 57 | 65 | 68 | 68 | 62 | 63 | 65 | 67 | 60 | 47 |
| **HAWAII** | | | | | | | | | | | | |
| Hilo | 48 | 42 | 41 | 34 | 31 | 41 | 44 | 38 | 42 | 41 | 34 | 36 |
| Honolulu | 62 | 64 | 60 | 62 | 64 | 66 | 67 | 70 | 70 | 68 | 63 | 60 |
| Lihue | 48 | 48 | 48 | 46 | 51 | 60 | 58 | 59 | 67 | 58 | 51 | 49 |
| **IDAHO** | | | | | | | | | | | | |
| Boise | 40 | 48 | 59 | 67 | 68 | 75 | 89 | 86 | 81 | 66 | 46 | 37 |
| Pocatello | 37 | 47 | 58 | 64 | 66 | 72 | 82 | 81 | 78 | 66 | 48 | 36 |
| **ILLINOIS** | | | | | | | | | | | | |
| Cairo | 46 | 53 | 59 | 65 | 71 | 77 | 82 | 79 | 75 | 73 | 56 | 46 |
| Chicago | 44 | 49 | 53 | 56 | 63 | 69 | 73 | 70 | 65 | 61 | 47 | 41 |
| Springfield | 47 | 51 | 54 | 58 | 64 | 69 | 76 | 72 | 73 | 64 | 53 | 45 |
| **INDIANA** | | | | | | | | | | | | |
| Evansville | 42 | 49 | 55 | 61 | 67 | 73 | 78 | 76 | 73 | 67 | 52 | 42 |
| Fort Wayne | 38 | 44 | 51 | 55 | 62 | 69 | 74 | 69 | 64 | 58 | 41 | 38 |
| Indianapolis | 41 | 47 | 49 | 55 | 62 | 68 | 74 | 70 | 68 | 64 | 48 | 39 |
| **IOWA** | | | | | | | | | | | | |
| Des Moines | 56 | 56 | 56 | 59 | 62 | 66 | 75 | 70 | 64 | 64 | 53 | 48 |
| Dubuque | 48 | 52 | 52 | 58 | 60 | 63 | 73 | 67 | 61 | 55 | 44 | 40 |
| Sioux City | 55 | 58 | 58 | 59 | 63 | 67 | 75 | 72 | 67 | 65 | 53 | 50 |
| **KANSAS** | | | | | | | | | | | | |
| Concordia | 60 | 60 | 62 | 63 | 65 | 73 | 79 | 76 | 72 | 70 | 64 | 58 |
| Dodge City | 67 | 66 | 68 | 68 | 68 | 74 | 78 | 78 | 76 | 75 | 70 | 67 |
| Wichita | 61 | 63 | 64 | 64 | 66 | 73 | 80 | 77 | 73 | 69 | 67 | 59 |
| **KENTUCKY** | | | | | | | | | | | | |
| Louisville | 41 | 47 | 52 | 57 | 64 | 68 | 72 | 69 | 68 | 64 | 51 | 39 |

| STATE/PROVINCE & CITY | JAN | FEB | MAR | APR | MAY | JUNE | JULY | AUG | SEPT | OCT | NOV | DEC |
|---|---|---|---|---|---|---|---|---|---|---|---|---|
| **LOUISIANA** | | | | | | | | | | | | |
| New Orleans | 49 | 50 | 57 | 63 | 66 | 64 | 58 | 60 | 64 | 70 | 60 | 46 |
| Shreveport | 48 | 54 | 58 | 60 | 69 | 78 | 79 | 80 | 79 | 77 | 65 | 60 |
| **MAINE** | | | | | | | | | | | | |
| Eastport | 45 | 51 | 52 | 52 | 51 | 53 | 55 | 57 | 54 | 50 | 37 | 40 |
| Portland | 55 | 59 | 58 | 57 | 57 | 64 | 66 | 63 | 62 | 59 | 51 | 49 |
| **MASSACHUSETTS** | | | | | | | | | | | | |
| Boston | 47 | 56 | 57 | 56 | 59 | 62 | 64 | 63 | 61 | 58 | 48 | 48 |
| **MICHIGAN** | | | | | | | | | | | | |
| Alpena | 29 | 43 | 52 | 56 | 59 | 64 | 70 | 64 | 52 | 44 | 24 | 22 |
| Detroit | 34 | 42 | 48 | 52 | 58 | 65 | 69 | 66 | 61 | 54 | 35 | 29 |
| Grand Rapids | 26 | 37 | 48 | 54 | 60 | 66 | 72 | 67 | 58 | 50 | 31 | 22 |
| Marquette | 31 | 40 | 47 | 52 | 53 | 56 | 63 | 57 | 47 | 38 | 24 | 24 |
| Sault Ste. Marie | 28 | 44 | 50 | 54 | 54 | 59 | 63 | 58 | 45 | 36 | 21 | 22 |
| **MINNESOTA** | | | | | | | | | | | | |
| Duluth | 47 | 55 | 60 | 58 | 58 | 60 | 68 | 63 | 53 | 47 | 36 | 40 |
| Minneapolis | 49 | 54 | 55 | 57 | 60 | 64 | 72 | 69 | 60 | 54 | 40 | 40 |
| **MISSISSIPPI** | | | | | | | | | | | | |
| Vicksburg | 46 | 50 | 57 | 64 | 69 | 73 | 69 | 72 | 74 | 71 | 60 | 45 |
| **MISSOURI** | | | | | | | | | | | | |
| Kansas City | 55 | 57 | 59 | 60 | 64 | 70 | 76 | 73 | 70 | 67 | 59 | 52 |
| St. Louis | 48 | 49 | 56 | 59 | 64 | 68 | 72 | 68 | 67 | 65 | 54 | 44 |
| Springfield | 48 | 54 | 57 | 60 | 63 | 69 | 77 | 72 | 71 | 65 | 58 | 48 |
| **MONTANA** | | | | | | | | | | | | |
| Havre | 49 | 58 | 61 | 63 | 63 | 65 | 78 | 75 | 64 | 57 | 48 | 48 |
| Helena | 46 | 55 | 58 | 60 | 59 | 63 | 77 | 74 | 63 | 57 | 48 | 43 |
| Kalispell | 28 | 40 | 49 | 57 | 58 | 60 | 77 | 73 | 61 | 50 | 28 | 20 |
| **NEBRASKA** | | | | | | | | | | | | |
| Lincoln | 57 | 59 | 60 | 60 | 63 | 69 | 76 | 71 | 67 | 66 | 59 | 55 |
| North Platte | 63 | 63 | 64 | 62 | 64 | 72 | 78 | 74 | 72 | 70 | 62 | 58 |
| **NEVADA** | | | | | | | | | | | | |
| Ely | 61 | 64 | 68 | 65 | 67 | 79 | 79 | 81 | 81 | 73 | 67 | 62 |
| Las Vegas | 74 | 77 | 78 | 81 | 85 | 91 | 84 | 86 | 92 | 84 | 83 | 75 |
| Reno | 59 | 64 | 69 | 75 | 77 | 82 | 90 | 89 | 86 | 76 | 68 | 56 |
| Winnemucca | 52 | 60 | 64 | 70 | 76 | 83 | 90 | 90 | 86 | 75 | 62 | 53 |
| **NEW HAMPSHIRE** | | | | | | | | | | | | |
| Concord | 48 | 53 | 55 | 53 | 51 | 56 | 57 | 58 | 55 | 50 | 43 | 43 |
| **NEW JERSEY** | | | | | | | | | | | | |
| Atlantic City | 51 | 57 | 58 | 59 | 62 | 65 | 67 | 66 | 65 | 54 | 58 | 52 |
| **NEW MEXICO** | | | | | | | | | | | | |
| Albuquerque | 70 | 72 | 72 | 76 | 79 | 84 | 76 | 75 | 81 | 85 | 79 | 70 |

| STATE/PROVINCE & CITY | JAN | FEB | MAR | APR | MAY | JUNE | JULY | AUG | SEPT | OCT | NOV | DEC |
|---|---|---|---|---|---|---|---|---|---|---|---|---|
| Roswell | 69 | 72 | 75 | 77 | 76 | 80 | 76 | 75 | 74 | 74 | 74 | 69 |
| **NEW YORK** | | | | | | | | | | | | |
| Albany | 43 | 51 | 53 | 53 | 57 | 62 | 63 | 61 | 58 | 54 | 39 | 38 |
| Binghamton | 31 | 39 | 41 | 44 | 50 | 56 | 54 | 51 | 47 | 43 | 29 | 26 |
| Buffalo | 32 | 41 | 49 | 51 | 59 | 67 | 70 | 67 | 60 | 51 | 31 | 28 |
| Canton | 37 | 47 | 50 | 48 | 54 | 61 | 63 | 61 | 54 | 45 | 30 | 31 |
| New York | 49 | 56 | 57 | 59 | 62 | 65 | 66 | 64 | 64 | 61 | 53 | 50 |
| Syracuse | 31 | 38 | 45 | 50 | 58 | 64 | 67 | 63 | 58 | 47 | 29 | 26 |
| **NORTH CAROLINA** | | | | | | | | | | | | |
| Asheville | 48 | 53 | 56 | 61 | 64 | 63 | 59 | 59 | 62 | 64 | 59 | 48 |
| Raleigh | 50 | 56 | 59 | 64 | 67 | 65 | 62 | 62 | 63 | 64 | 62 | 52 |
| **NORTH DAKOTA** | | | | | | | | | | | | |
| Bismark | 52 | 58 | 56 | 57 | 58 | 61 | 73 | 69 | 62 | 59 | 49 | 48 |
| Devils Lake | 53 | 60 | 59 | 60 | 59 | 62 | 71 | 67 | 59 | 56 | 44 | 45 |
| Fargo | 47 | 55 | 56 | 58 | 62 | 63 | 73 | 69 | 60 | 57 | 39 | 46 |
| Williston | 51 | 59 | 60 | 63 | 66 | 66 | 78 | 75 | 65 | 60 | 48 | 48 |
| **OHIO** | | | | | | | | | | | | |
| Cincinnati | 41 | 46 | 52 | 56 | 62 | 69 | 72 | 68 | 68 | 60 | 46 | 39 |
| Cleveland | 29 | 36 | 45 | 52 | 61 | 67 | 71 | 68 | 62 | 54 | 32 | 25 |
| Columbus | 36 | 44 | 49 | 54 | 63 | 68 | 71 | 68 | 66 | 60 | 44 | 35 |
| **OKLAHOMA** | | | | | | | | | | | | |
| Oklahoma City | 57 | 60 | 63 | 64 | 65 | 74 | 75 | 78 | 74 | 68 | 64 | 57 |
| **OREGON** | | | | | | | | | | | | |
| Baker | 41 | 49 | 56 | 61 | 63 | 67 | 83 | 81 | 74 | 62 | 46 | 37 |
| Portland | 27 | 34 | 41 | 49 | 52 | 55 | 70 | 65 | 55 | 42 | 29 | 23 |
| Roseburg | 24 | 32 | 40 | 51 | 57 | 59 | 79 | 77 | 65 | 42 | 28 | 28 |
| **PENNSYLVANIA** | | | | | | | | | | | | |
| Harrisburg | 43 | 52 | 55 | 57 | 61 | 63 | 68 | 63 | 62 | 58 | 47 | 43 |
| Philadelphia | 45 | 56 | 57 | 58 | 61 | 62 | 64 | 61 | 62 | 61 | 53 | 49 |
| Pittsburgh | 32 | 38 | 45 | 50 | 57 | 62 | 64 | 61 | 62 | 54 | 39 | 30 |
| **RHODE ISLAND** | | | | | | | | | | | | |
| Block Island | 45 | 54 | 47 | 56 | 58 | 60 | 62 | 62 | 60 | 59 | 50 | 44 |
| **SOUTH CAROLINA** | | | | | | | | | | | | |
| Charleston | 58 | 60 | 65 | 72 | 73 | 70 | 66 | 66 | 67 | 68 | 68 | 57 |
| Columbia | 53 | 57 | 62 | 68 | 69 | 68 | 63 | 65 | 64 | 68 | 64 | 51 |
| **SOUTH DAKOTA** | | | | | | | | | | | | |
| Huron | 55 | 62 | 60 | 62 | 65 | 68 | 76 | 72 | 68 | 61 | 52 | 48 |
| Rapid City | 58 | 62 | 63 | 62 | 61 | 66 | 73 | 73 | 69 | 66 | 58 | 54 |
| **TENNESSEE** | | | | | | | | | | | | |
| Knoxville | 42 | 49 | 53 | 59 | 64 | 66 | 64 | 59 | 64 | 64 | 53 | 41 |
| Memphis | 44 | 51 | 57 | 64 | 68 | 74 | 73 | 74 | 70 | 69 | 58 | 45 |
| Nashville | 42 | 47 | 54 | 60 | 65 | 69 | 69 | 68 | 69 | 65 | 55 | 42 |

| STATE/PROVINCE & CITY | JAN | FEB | MAR | APR | MAY | JUNE | JULY | AUG | SEPT | OCT | NOV | DEC |
|---|---|---|---|---|---|---|---|---|---|---|---|---|
| **TEXAS** | | | | | | | | | | | | |
| Abilene | 64 | 68 | 73 | 66 | 73 | 86 | 83 | 85 | 73 | 71 | 72 | 66 |
| Amarillo | 71 | 71 | 75 | 75 | 75 | 82 | 81 | 81 | 79 | 76 | 76 | 70 |
| Austin | 46 | 50 | 57 | 60 | 62 | 72 | 76 | 79 | 70 | 70 | 57 | 49 |
| Brownsville | 44 | 49 | 51 | 57 | 65 | 73 | 78 | 78 | 67 | 70 | 54 | 44 |
| Del Rio | 53 | 55 | 61 | 63 | 60 | 66 | 75 | 80 | 69 | 66 | 58 | 52 |
| El Paso | 74 | 77 | 81 | 85 | 87 | 87 | 78 | 78 | 80 | 82 | 80 | 73 |
| Fort Worth | 56 | 57 | 65 | 66 | 67 | 75 | 78 | 78 | 74 | 70 | 63 | 58 |
| Galveston | 50 | 50 | 55 | 61 | 69 | 76 | 72 | 71 | 70 | 74 | 62 | 49 |
| San Antonio | 48 | 51 | 56 | 58 | 60 | 69 | 74 | 75 | 69 | 67 | 55 | 49 |
| **UTAH** | | | | | | | | | | | | |
| Salt Lake City | 48 | 53 | 61 | 68 | 73 | 78 | 82 | 82 | 84 | 75 | 56 | 49 |
| **VERMONT** | | | | | | | | | | | | |
| Burlington | 34 | 43 | 48 | 47 | 53 | 59 | 62 | 59 | 51 | 43 | 25 | 24 |
| **VIRGINIA** | | | | | | | | | | | | |
| Norfolk | 50 | 57 | 60 | 63 | 67 | 66 | 66 | 66 | 63 | 61 | 60 | 51 |
| Richmond | 49 | 55 | 59 | 63 | 67 | 66 | 65 | 62 | 63 | 64 | 58 | 50 |
| **WASHINGTON** | | | | | | | | | | | | |
| North Head | 26 | 37 | 42 | 48 | 48 | 48 | 50 | 46 | 48 | 41 | 31 | 27 |
| Seattle | 27 | 34 | 42 | 48 | 53 | 48 | 62 | 56 | 53 | 36 | 28 | 24 |
| Spokane | 26 | 41 | 53 | 63 | 64 | 68 | 82 | 79 | 68 | 53 | 28 | 22 |
| Tatoosh Island | 26 | 36 | 39 | 45 | 47 | 46 | 48 | 44 | 47 | 38 | 26 | 23 |
| Walla Walla | 24 | 35 | 51 | 63 | 67 | 72 | 86 | 84 | 72 | 59 | 33 | 20 |
| Yakima | 34 | 49 | 62 | 70 | 72 | 74 | 86 | 86 | 74 | 61 | 38 | 29 |
| **WEST VIRGINIA** | | | | | | | | | | | | |
| Elkins | 33 | 37 | 42 | 47 | 55 | 55 | 56 | 53 | 55 | 51 | 41 | 33 |
| Parkersburg | 30 | 36 | 42 | 49 | 56 | 60 | 63 | 60 | 60 | 53 | 37 | 29 |
| **WISCONSIN** | | | | | | | | | | | | |
| Green Bay | 44 | 51 | 55 | 56 | 58 | 64 | 70 | 65 | 58 | 52 | 40 | 40 |
| Madison | 44 | 49 | 52 | 53 | 58 | 64 | 70 | 66 | 60 | 58 | 41 | 38 |
| Milwaukee | 44 | 48 | 53 | 56 | 60 | 65 | 73 | 67 | 62 | 56 | 44 | 39 |
| **WYOMING** | | | | | | | | | | | | |
| Cheyenne | 65 | 66 | 64 | 61 | 59 | 68 | 70 | 68 | 69 | 69 | 65 | 63 |
| Lander | 66 | 70 | 71 | 66 | 65 | 74 | 76 | 75 | 72 | 67 | 61 | 62 |
| Sheridan | 56 | 61 | 62 | 61 | 61 | 67 | 76 | 74 | 67 | 60 | 53 | 52 |
| Yellowstone Park | 39 | 51 | 55 | 57 | 56 | 63 | 73 | 71 | 65 | 57 | 45 | 38 |
| **ALBERTA** | | | | | | | | | | | | |
| Edmonton | 35 | 43 | 45 | 53 | 52 | 49 | 61 | 58 | 49 | 48 | 39 | 33 |
| **BRITISH COLUMBIA** | | | | | | | | | | | | |
| Prince George | 22 | 31 | 36 | 44 | 50 | 47 | 52 | 53 | 43 | 31 | 22 | 18 |
| Vancouver | 16 | 26 | 30 | 41 | 47 | 43 | 56 | 56 | 46 | 32 | 19 | 13 |

| STATE/PROVINCE & CITY | JAN | FEB | MAR | APR | MAY | JUNE | JULY | AUG | SEPT | OCT | NOV | DEC |
|---|---|---|---|---|---|---|---|---|---|---|---|---|
| **MANITOBA** | | | | | | | | | | | | |
| WInnipeg | 38 | 47 | 45 | 50 | 51 | 51 | 63 | 60 | 48 | 46 | 30 | 32 |
| **NEWFOUNDLAND** | | | | | | | | | | | | |
| Gander | 26 | 29 | 29 | 28 | 32 | 33 | 41 | 40 | 38 | 33 | 23 | 23 |
| **NOVA SCOTIA** | | | | | | | | | | | | |
| Halifax | 34 | 39 | 40 | 38 | 44 | 46 | 51 | 50 | 45 | 44 | 31 | 33 |
| **ONTARIO** | | | | | | | | | | | | |
| Kapuskasing | 27 | 36 | 37 | 41 | 41 | 43 | 48 | 45 | 33 | 27 | 16 | 21 |
| Toronto | 27 | 35 | 38 | 42 | 48 | 56 | 61 | 60 | 53 | 45 | 29 | 27 |
| **QUEBEC** | | | | | | | | | | | | |
| Montreal | 29 | 36 | 40 | 41 | 44 | 47 | 51 | 51 | 45 | 37 | 24 | 28 |
| **SASKATCHEWAN** | | | | | | | | | | | | |
| Regina | 37 | 41 | 41 | 52 | 55 | 51 | 67 | 63 | 52 | 51 | 35 | 34 |

# RESOURCE GUIDE

THIS GUIDE *provides sources of additional information for all major topics covered in the book. Here you will find important books and articles, videos, magazines, newsletters, as well as organizations and suppliers. For Web sites and important links to on-line information, or to contact me, visit my Web site, www.chelseagreen.com/Chiras.*

## PASSIVE SOLAR HEATING AND INTEGRATED DESIGN (Chapters 1, 3, and 7)

### PUBLICATIONS

Aulisi, Susan, and Doug McGilvray. *House Warming*. Edinburg, N.Y.: Adirondack Alternate Energy, 1983. Overview of passive solar heating with some interesting design ideas.

Chiras, Daniel D. "Build a Solar Home and Let the Sun Shine In," *Mother Earth News*, August/September 2002, pp. 74–81. A survey of passive solar design principles, also showing the economics of passive solar heating.

Chiras, Daniel D., ed. "Solar Solutions," *The Last Straw* 36 (Winter 2001). A collection of over a dozen articles, many by the author, on passive solar heating, integrated design, thermal mass, and more.

Chiras, Dan. "Learning from Mistakes of the Past," *The Last Straw* 36 (Winter 2001), 15-16. Describes common errors in passive solar design.

Cole, Nancy, and P.J. Skerrett. *Renewables Are Ready: People Creating Renewable Energy Solutions*. White River Junction, Vt.: Chelsea Green, 1995. Contains numerous interesting case studies showing how people have applied various solar technologies, including passive solar.

Crosbie, Michael. J., ed. *The Passive Solar Design and Construction Handbook*. New York: John Wiley and Sons, 1997. A pricey and fairly technical manual on passive solar homes. Contains detailed drawings and case studies.

Crowther, Richard I. *Affordable Passive Solar Homes: Low-Cost Compact Designs*. Denver, Co.: SciTech Publishing, 1984. Contains some valuable background information on passive solar design and numerous designs for passive solar homes.

Energy Division, North Carolina Department of Commerce. *Solar Homes for North Carolina: A Guide to Building and Planning Solar Homes*. Raleigh, N.C.: North Carolina Solar Center, 1999. Available on-line at the North Carolina Solar Center's Web site. (See Organizations.)

Freeman, Mark. *The Solar Home: How to Design and Build a House You Heat with the Sun.* Mechanicsburg, Pa.: Stackpole Books, 1994. Fairly useful introduction, although it contains more information on general building than passive solar design and construction.

Jones, Leonard D. "Thermal Mass in Passive Applications," *The Last Straw* 36 (Winter 2001), 10-12. A good, fairly detailed introduction to the function of thermal mass.

Kachadorian, James. *The Passive Solar House.* White River Junction, Vt.: Chelsea Green, 1997. Presents a lot of good information on passive solar heating and an interesting design that has reportedly been fairly successful in cold climates.

Kubsch, E. *Homeowner's Guide to Free Heat: Cutting Your Heating Bills Over 50%.* Sheridan, Wy.: Sunstore Farms, 1991. A self-published book with lots of good, basic information.

Miller, Burke. *Solar Energy: Today's Technologies for a Sustainable Future.* Boulder, Co.: American Solar Energy Society, 1997. An extremely valuable resource with numerous case studies showing how passive solar heating can be used in different climates, even some fairly solar-deprived places.

Olson, Ken, and Joe Schwartz. "Home Sweet Solar Home," *Home Power* 90 (Aug./Sept. 2002), 86–94.

Passive Solar Industries Council. *Passive Solar Design Strategies: Guidelines for Home Builders.* Washington, D.C.: PSIC, undated. Extremely useful book with worksheets for calculating a house's energy demand, the amount of back-up heat required, the temperature swing one can expect given the amount of thermal mass you've installed, and the estimated cooling load. You can order a copy from the Sustainable Buildings Industry Council (formerly the PSIC) with detailed information for your state, so you can design a home to meet the requirements of your site.

Potts, Michael. *The New Independent Home: People and Houses that Harvest the Sun, Wind, and Water.* White River Junction, Vt.: Chelsea Green, 1999. Delightfully readable book with lots of good information.

Reynolds, Michael. *Comfort in Any Climate.* Taos, N.M.: Solar Survival Press, 1990. A brief, but informative treatise on passive heating and cooling.

Sklar, Scott, and Kenneth Sheinkopf. *Consumer Guide to Solar Energy: More Ways to Reduce Your Energy Bills and Save the Environment.* Chicago, Il.: Bonus Books, 1995. Delightfully written introduction to many different solar applications, including passive solar heating.

Solar Survival Architecture. "Thermal Mass vs. Insulation." *Earthship Chronicles.* Taos, N.M.: Solar Survival Architecture, 1998. Basic treatise on passive solar heating and cooling.

Sustainable Buildings Industry Council. *Designing Low-Energy Buildings: Passive Solar Strategies* and *Energy-10 Software.* SBIC, 1996. A superb resource! This book of design guidelines and the *Energy-10* software that comes with it enables builders to analyze the energy and cost savings in building designs. Helps permit region-specific design.

Taylor, John S. *Shelter Sketchbook: Timeless Building Solutions.* White River Junction, Vt.: Chelsea Green, 1983. Pictorial history of building that will open your eyes to intriguing design solutions to achieve comfort, efficiency, convenience, and beauty.

Van Dresser, Peter. *Passive Solar House Basics.* Santa Fe, N.M.: Ancient City Press, 1996. This brief book provides basics on passive solar design and construction, primarily of adobe homes. Contains sample house plans, ideas for solar water heating, and much more.

**VIDEOS**

*Buildings for a Sustainable America.* A concise overview of passive solar buildings and their benefits. Available from the Sustainable Buildings Industry Council (SBIC),

1331 H Street NW, Suite 1000, Washington, D.C. 20005. Tel: (202) 628-7400. Web site: www.sbicouncil.org.

*The Solar-Powered Home with Rob Roy.* An 84-minute video that examines basic principles, components, set-up, and system planning for an off-grid home featuring tips from America's leading experts in the field of home power. Can be purchased from the Earthwood Building School at 366 Murtagh Hill Road, West Chazy, N.Y. 12992. Tel: (518) 493-7744. Web site: www.interlog.com/~ewood.

## MAGAZINES AND NEWSLETTERS

*Backwoods Home Magazine.* Publishes articles on all aspects of self-reliant living, including renewable energy strategies such as solar. P.O. Box 712, Gold Beach, OR 97444. Tel: (800) 835-2418. Web site: www.backhome.com.

*The CADDET Renewable Energy Newsletter.* Quarterly magazine published by the CADDET Centre for Renewable Energy, 168 Harwell, Oxfordshire OX11 ORA, United Kingdom. Tel: +44 123335 432968.

*Earth Quarterly* (formerly *Dry Country News*). A new magazine devoted to living close to, and in harmony with, nature. Covers all aspects of natural life including homebuilding and renewable energy. Box 23-J, Radium Springs, N.M. 88054. Tel: (505) 526-1853.Web site: www.zianet.com/ earth.

*EREN Network News.* Newsletter of the Department of Energy's Energy-Efficiency and Renewable Energy Network. See listing under organizations.

*Home Energy Magazine.* Great resource for those who want to learn more about ways to save energy in conventional home construction. 2124 Kittredge Street, No. 95, Berkeley, CA 94704.

*Home Power.* Publishes numerous articles on PVs, wind energy, microhydroelectric, and occasionally an article or two on passive solar heating and cooling. P.O. Box 520, Ashland, OR 97520. Tel: (800) 707-6585. Web site: www.homepower.com

*Inside and Out.* Newsletter of the Sustainable Buildings Industry Council. See their listing under organizations.

*The Last Straw.* This journal publishes articles on natural building and features articles on passive solar heating and cooling. Contact them at: TLS, HC 66, Box 119, Hillsboro, NM 88042. Tel: (505) 895-5400. Web site: www.strawhomes.com.

*Mother Earth News.* Publishes numerous articles on renewable energy and related topics. Ogden Publications, 105 S.W. 42nd St., Topeka, KS 66609. Tel: (785) 274-3400. Web site: www. mother earthnews.com.

*National Renewable Energy Lab Now.* Check out their newsletter on line at: www.nrel.gov.

*Solar Today.* This magazine published by the American Solar Energy Society contains a wealth of information on passive solar, solar thermal, photovoltaics, hydrogen, and other topics. Also lists names of engineers, builders, and installers and lists workshops and conferences. ASES, 2400 Central Ave., Suite G-1, Boulder, CO 80301. Tel: (303) 443-3130. Web site: www.solartoday.org.

## ORGANIZATIONS

**American Solar Energy Society.** Publishes *Solar Today* magazine and sponsors an annual national meeting. Also publishes an on-line catalogue of publications and sponsors the National Tour of Solar Homes. Contact this organization to find out about an ASES chapter in your area. 2400 Central Avenue, Suite G-1, Boulder, CO 80301. Web site: www.ases.org/solar/.

**Center for Building Science,** Lawrence Berkeley National Laboratory's Center for Building Science works to develop and commercialize energy-efficient technologies and to document ways of improving energy efficiency of homes and other buildings while protecting air quality. Web site: http://eande.lbl.gov/CBS/ CBS.html.

**Center for Renewable Energy and Sustainable Technologies (CREST).** Nonprofit organization dedicated to renewable energy, energy efficiency, and sustainable living. CREST, 1612 K St. NW, Suite 410, Washington, DC 20006. Tel: (202) 293-2898. Web site: http://solstice.crest.org.

**El Paso Solar Energy Association.** Active in solar energy, especially passive solar design and construction. P.O. Box 26384, El Paso, TX 79926.

**Energy Efficiency and Renewable Energy Clearinghouse.** Great source of a variety of useful information on renewable energy. P.O. Box 3048, Merrifield, VA 22116. Tel: (800) 363-3732.

**Florida Solar Energy Center.** A research institute of the University of Central Florida. Research and education on passive solar, cooling, and photovoltaics. FSEC, 1679 Clearlake Road, Cocoa, FL 32922. Tel: (321) 638-1000. Web site: http://www.fsec.ucf.edu.

**Midwest Renewable Energy Association.** Actively promotes solar energy and offers valuable workshops. P.O. Box 249, Amherst, WI 54406. Tel: (715) 824-5166. Web site: www.the-mrea.org.

**National Renewable Energy Lab.** Center for Buildings and Thermal Systems. Key players in research and education on energy efficiency and passive solar heating and cooling. NREL, 1617 Cole Blvd.,

Golden, CO 80401. Tel: (303) 384-7349. Web site: www.nrel.gov/buildings/highperformance.

**North Carolina Solar Center.** Offers workshops, tours, publications, and much more. Address: Box 7401, Raleigh, NC 27695. Tel: (919) 515-3480. Web site: http://www.ncsc.ncsu.edu.

**Renewable Energy Training and Education Center.** Offers hands-on training and certification courses in U.S. and abroad for those interested in becoming certified in solar installation. U.S. 1679 Clearlake Road, Cocoa, FL 32922. Tel: (407) 638-1007.

**Solar Energy International.** Offers a wide range of workshops on solar energy, wind energy, and natural building. Contact them at P.O. Box 715, Carbondale, CO 81623. Tel: (970) 963-8855. Web site: www.solarenergy.org.

**Sustainable Buildings Industries Council.** This organization has a terrific Web site with information on workshops, books and publications, and links to many other international, national, and state solar energy organizations. Publishes a newsletter, *Buildings Inside and Out.* SBIC, 331 H. Street NW, Suite 1000, Washington, DC 20005. Tel: (202) 628-7400. Web site: www.psic.org/.

## ENERGY-EFFICIENT DESIGN AND CONSTRUCTION (Chapter 2)

### PUBLICATIONS

Chiras, Dan. "Minimize the Digging: Frost-Protected Shallow Foundations," *The Last Straw* 38 (Summer 2002), p. 10. A brief treatise on frost-protected shallow foundations.

————. "Retrofitting a Foundation for Energy Efficiency," *The Last Straw* 38 (Summer 2002), p. 10. Describes ways to retrofit foundations to reduce heat loss.

Carmody, John, Stephen Selkowitz, and Lisa Heschong. *Residential Windows: A Guide to New Technologies and Energy Performance.* New York: Norton, 1996. Extremely important reading for all passive solar home designers.

Fine Homebuilding. *The Best of Fine Homebuilding: Energy-Efficient Building.* Newtown, Ct.: Taunton Press, 1999. A collection of detailed, somewhat technical articles on a wide assortment of topics related to energy efficiency including insulation, energy-saving details, windows, housewraps, skylights, and heating systems.

Lstiburek, Joe, and Besty Pettit. *EEBA Builder's Guide—Cold Climate.* Minneapolis: Energy Efficient Building Association, 1999. Superb resource for advice on building in cold climates.

———. *EEBA Builder's Guide—Mixed Humid Climate*. Minneapolis: Energy Efficient Building Association, 1999. Superb resource for advice on this climate.

———. *EEBA Builder's Guide—Hot-Arid Climate*. Minneapolis: EEBA, 1999. Superb resource for advice on building in hot arid climates.

Magwood, Chris, ed. "Roofs and Foundations," *The Last Straw* 38 (September 2002). An excellent resource for those who want to learn about energy- and material-efficient foundations.

National Association of Home Builders Research Center. *Design Guide for Frost-Protected Shallow Foundations*. Upper Marlboro, Md.: NAHB Research Center, 1996. Also available on-line.

Oehler, Mike. *The $50 and Up Underground House Book: How to Design and Build Underground*. A very popular book for those who want to live inexpensively off the beaten path.

Roy, Rob. *The Complete Book of Underground Houses: How to Build a Low-Cost Home*. New York: Sterling, 1994. A revision of a 1979 best-seller with new information on earth-sheltered homes. Can be purchased from the Earthwood Building School (listed earlier).

Sikora, Jeannie L. *Profit from Building Green: Award Winning Tips to Build Energy Efficient Homes*. Washington, D.C.: BuilderBooks, 2002. A brief, but informative overview of energy-conservation strategies.

Wells, Malcolm. *The Earth-Sheltered House: An Architect's Sketchbook*. White River Junction, Vt.: Chelsea Green, 1998. Although you won't find a ton of information on earth-sheltered housing in this book, you will be regaled with lots of inspiring designs that will help you see the potential of this design strategy.

Wilson, Alex, Jennifer Thorne, and John Morrill. *Consumer Guide to Home Energy Savings, 7th ed.* Washington, D.C.: American Council for an Energy-Efficient Economy, 1999. Excellent book, full of information on energy-saving appliances.

Yost, Harry. *Home Insulation: Do It Yourself and Save as Much as 40%*. Pownal, Vt.: Storey Communications, 1991. Extremely useful book for anyone building his or her own home.

## ORGANIZATIONS

**American Council for an Energy-Efficient Economy.** Numerous excellent publications on energy efficiency, including *Consumer Guide to Home Energy Savings*. 1001 Connecticut Avenue NW, Suite 801, Washington, DC 20036. Tel: (202) 429-0063. Web site: www.aceee.org.

**Building America Program.** Leaders in promoting energy efficiency and renewable energy to achieve zero-energy buildings. U.S. Department of Energy. Office of Building Systems, EE-41, 1000 Independence Avenue SW, Washington, DC 20585. Tel: (202) 586-9472.

**Cellulose Insulation Manufacturers Association.** Your place to "shop" for information on cellulose insulation. 133 S. Keowee St., Dayton, OH 45402. Tel: (937) 222-2462. Web site: www.cellulose.org.

**Energy Efficiency and Renewable Energy Clearinghouse.** Great source of a variety of useful information on energy efficiency. P.O. Box 3048, Merrifield, VA 22116. Tel: (800) 363-3732.

**Energy Efficient Building Association.** Offers conferences, workshops, publications and an on-line bookstore. 490 Concordia Ave., P.O. Box 22307, Eagen, MN 55122. Tel: (952) 881-1098.

**Insulating Concrete Forms Association.** A great place to begin your research on ICFs. 1807 Glenview Rd., Suite 203, Glenview, IL 60025. Tel: (847) 657-9730. Web site: www.forms.org.

**National Fenestration Rating Council.** For information on energy efficiency of windows. 8484 Georgia Ave., Suite 320, Silver Springs, MD 20910. Tel: (301) 589-1776. Web site: www.nfrc.org.

**National Insulation Association.** Offers a wide range of information on different types of insulation. 99 Canal Center Plaza, Suite 222, Alexandria, VA 22314. Tel: (703) 683-6422. Web site: www.insulation.org.

## BACK-UP HEATING (Chapter 4)

### Radiant Floor and Baseboard Hot Water Systems

#### PUBLICATIONS

Fust, Art. "A Simple Warm Floor Heating System," *The Last Straw* 32 (Winter 2000), 25-26. Contains much useful information.

Grahl, Christine L. "The Radiant Flooring Revolution," *Environmental Design and Construction* (January/February 2000), 38-40. Superb introduction to radiant-floor heating.

Hyatt, Rod. "Hydronic Heating on Renewable Energy," *Home Power* 79 (October/November 2000), 36-42. Provides a lot of practical advice on building your own radiant-floor heating system and powering it with photovoltaic panels.

Siegenthaler, John. "Hydronic Radiant-Floor Heating," *Fine Homebuilding* (October/November 1996), 58-63. Extremely useful reference. Well written, thorough, and well illustrated.

———. *Modern Hydronic Heating*. Albany, N.Y.: Delmar Publish-

ers, 1995. Everything you would ever want to know about hydronic heating.

Wilson, Alex. "Radiant-Floor Heating: When It Does—and Doesn't—Make Sense," *Environmental Building News* 11 (January 2002), 1, 9-14. Valuable reading.

#### ORGANIZATIONS

**Radiant Panel Association.** Professional organization consisting of radiant heating and cooling contractors, wholesalers, manufacturers, and professionals. 1433 West 29th Street, Loveland, CO 80539. Tel: (970) 613-0100. Web site: www.radiantpanelassociation.org.

### Forced-Air Heating, Furnaces and Boilers

#### PUBLICATIONS

Fine Homebuilding. *Energy-Efficient Building*. Newtown, Ct: Taunton Press, 1999. Contains a collection of extremely useful articles on heating systems.

O'Connell, John, and Bruce Harley. "Choosing Ductwork," *Fine Homebuilding* 110 (June/July 1997, 98-101. Essential reading for anyone interested in installing a forced-air heating system.

Wilson, Alex. "A Primer on Heating Systems," *Fine Homebuilding* 110 (February/March1997), 50-55. Superb overview of furnaces, boilers, and heating systems.

### Wall-Mounted Space Heaters

Consumer Product Safety Commission. For a wealth of information on space heaters, including safety precautions, contact Office of Information and Public Affairs, CPSC, Washington, D.C. 20207 or call their hotline at (800) 638-2772. Web site: www.cspc.gov.

### Heat Pumps

#### PUBLICATIONS

Malin, Nadav, and Alex Wilson. "Ground-Source Heat Pumps: Are They Green?" *Environmental Building News* 9 (July/August 2000), 1, 16-22. Detailed overview of the operation and pros and cons of ground-source heat pumps.

National Renewable Energy Lab. "Geothermal Heat Pumps," published on-line at http://www.eren.doe.gov/erec/factsheets/geo_heat-pumps.html. Great overview of GSHPs.

#### ORGANIZATIONS

**Geo-Heat Center,** Oregon Institute of Technology, 3201 Campus Dr., Klamath, OR 97601. Tel: (541) 885-1750. Web site: www.oit.osshe.edu/~geoheat/.

**Geothermal Heat Pump Consortium, Inc.** 701 Pennsylvania Ave, NW, Washington, DC 20004-2696. Tel: (888) 333-4472. Web site: www.ghpc.org.

**International Ground Source Heat Pump Association.** Provides a list of equipment manufacturers, installers by state, and numerous other resources for contractors, homeowners, students, and the general public. 490 Cordell South, Stillwater, OK 74078-8018. Tel: (405) 744-5175. Web site: www.igshpa.okstate.edu/.

**U.S. Department of Energy, Office of Geothermal Technologies.** Carries out research on GSHPs and works closely with industry to implement new ideas. EE-12, 1000 Independence Avenue, SW, Washington, DC 20585-0121. Tel: (202) 586-5340.

## Woodstoves and Masonry Heaters

### PUBLICATIONS

Barden, Albert A. AlbiCoreTM Construction Manual. Norridgewock, Me.: Maine Wood Heat Company, 1996. Detailed construction manual.

Barden, Albert A. *The Finnish Fireplace: Construction Manual.* Norridgewock, ME: Maine Wood Heat Company, Inc., 1984. The only complete English language primer on making masonry heaters. Available through the Maine Wood Heat Company (listed above).

———— and Keikki Hyytiainen. *Finnish Fireplaces: Heart of the Home.* Finland: Building Book Ltd., 1988. A valuable resource for anyone wanting to learn more about Finnish masonry stoves. Available through the Maine Wood Heat Company (listed above).

British Columbia Ministry of Environment, Land, and Parks. "Reducing Wood Stove Smoke: A Burning Issue," Sept.1994. Web site: www.env.gov.bc.ca/epd/epdqa/ar/particulates/rwssabi.html.

Gulland, John. "Woodstove Buyer's Guide," *Mother Earth News* (December/January 2002), 32-43. Superb overview of woodstoves with a useful table to help you select a model that meets your needs.

Johnson, Dave. *The Good Woodcutter's Guide: Chain Saws, Portable Sawmills, and Woodlots.* White River Junction, Vt.: Chelsea Green, 1998. A practical guide to felling trees and cutting fire wood safely.

Lyle, David. *The Book of Masonry Stoves: Rediscovering an Old Way of Warming.* White River Junction, Vt.: Chelsea Green, 1984. This book contains a wealth of information on the history, function, design, and construction of masonry stoves.

### ORGANIZATIONS

**Hearth, Patio, and Barbecue Association.** (Formerly the Hearth Products Association.) International trade association that promotes the interests of the hearth products industry. Offers lots of valuable information. 1601 North Kent Street, Suite 1001, Arlington, VA 22209. Web site: http://hpba.org.

**Masonry Heater Association of North America.** Publishes a valuable newsletter and has a Web site with links to dealers and masons who design and build masonry stoves. 1252 Stock Farm Road, Randolph, VT 05060. Tel: (802) 728-5896. Web site: www.mha-net.org.

**Wood Heat Organization.** Promotes safe, responsible use of wood for heating. Contact them at: 410 Bank Street, Suite 117, Ottawa, Ontario Canada K2P 1Y8. Web site: www.woodheat.org.

## PASSIVE COOLING (Chapter 6)

### PUBLICATIONS

Givoni, Baruch. *Passive and Low Energy Cooling of Buildings.* New York: John Wiley and Sons, 1994. A fairly technical book, but one of few resources on the subject.

## HEALTHY/GREEN BUILDING (Chapter 8)

### PUBLICATIONS

Borer, Pat, and Cindy Harris. *The Whole House Book: Ecological Building Design and Materials.* Powys, England: Centre for Alternative Technology Publications, 1998. Contains a wealth of information on building healthy, environmentally friendly homes.

Baker-Laporte, Paula, Erica Elliot, and John Banta. *Prescriptions for a Healthy House: A Practical Guide for Architects, Builders, and Homeowner*s. 2nd ed. Gabriola Island, B.C.: New Society Publishers, 2001. Superb resource with a great amount of useful information.

Bower, John. *The Healthy House: How to Buy One, How to Build One, How to Cure a Sick One.* 3rd ed., Bloomington, In.: The Healthy House Institute, 1997. A very detailed guide to all aspects of home construction. Worth its weight in gold.

——— and Lynn Marie Bower. *The Healthy House Answer Book: Answers to the 133 Most Commonly Asked Questions.* Bloomington, In.: The Healthy House Institute, 1997. Great resource for those who just want to learn the basics.

Chappell, Steve K., ed. *The Alternative Building Sourcebook.* Fox Maple Press: Brownfield, Me., 1998. Lists over 900 products and professional services in the area of natural and sustainable building.

Chiras, Daniel D. *The Natural House: A Complete Guide to Healthy, Energy-Efficient, Environmental Homes.* White River Jct., Vt.: Chelsea Green, 2000. A comprehensive survey of natural building with additional information on passive solar heating and cooling, green building materials, and other topics.

City of Austin Green Builder Program. *Sustainable Building Sourcebook.* Austin: City of Austin Green Builder Program. Excellent resource, available on-line at www.ci.austin.tx.us/greenbuilder/.

Davis, Andrew N. and Paul E. Schaffman. *The Home Environmental Sourcebook: 50 Environmental Hazards to Avoid When Buying, Selling, or Maintaining a Home.* New York: Henry Holt, 1996. Good overview of sources of health hazards in homes.

U.S. Environmental Protection Agency. *The Inside Story: A Guide to Indoor Air Quality.* Washington, D.C.: EPA, 1995. Very helpful on-line publication for those interested in learning more about indoor air quality issues and solutions. You can access it at www.epa.gov/iaq/insidest.html.

———. *Indoor Air Pollution: An Introduction for Health Professionals.* Washington, D.C.: EPA, 1994. A detailed guide on air pollution and health effects. Very valuable for diagnosing problems caused by indoor air pollution. Also contains an extensive bibliography of research papers on the subject. Available at: www.epa.gov/iaq/pubs/hpguide.html.

———. *Model Standards and Techniques for Control of Radon in New Residential Buildings.* Washington, D.C.: EPA, 1994. This on-line document provides detailed, fairly technical information on ways to prevent radon from becoming a problem in new construction. Available at: www.epa.gov/iaq/radon/pubs/newconst.html.

———. *A Citizen's Guide to Radon. The Guide to Protecting Yourself and Your Family from Radon.* 2nd ed. Washington, D.C.: EPA, 1992. A very basic on-line introduction to radon. Available at: www.epa.gov/iaq/radon/pubs/citguide.html.

———. *What You Should Know About Combustion Appliances and Indoor Air Quality.* Washington, D.C.: EPA, undated. A great little introduction to the effects of indoor air pollutants from combustion sources. Available at: www.epa/iaq/pubs/combust.html.

——— and the U.S. Consumer Product Safety Commission. *The Inside Story: A Guide to Indoor Air Quality.* EPA Document No. 402-K-93-007. Washington, D.C.: U.S. Government Printing Office, 1995.

Hermannsson, James. *Green Building Resource Guide.* Newtown, Ct: Taunton Press, 1997. A goldmine of information on environmentally friendly building materials. Reader beware: not all building materials in books such as this pass the sustainability test.

Holmes, Dwight, Larry Strain, Alex Wilson, and Sandra Leibowitz. *GreenSpec: The Environmental Building News Product*

*Directory and Guideline Specifications.* BuildingGreen, Inc.: Brattleboro, Vt., 1999. Guideline specifications make this an extremely valuable resource for commercial builders and architects.

Pearson, David. *The Natural House Catalog: Everything You Need to Create An Environmentally Friendly Home.* New York: Simon and Schuster, 1996. Contains a lot of information on building and furnishing a sustainable home, including a list of products and services.

Rousseau, David, and James Wasley. *Healthy by Design: Building and Remodeling Solutions for Creating Healthy Homes.* Point Roberts, Wa: Hartley and Marks Publishers, 1999. Great book with lots of useful information.

Spiegel, Ross, and Dru Meadows. *Green Building Materials: A Guide to Product Selection and Specification.* New York: John Wiley and Sons, 1999. The newest entry into the green building materials books. Looks like a great resource.

## MAGAZINES AND NEWSLETTERS

*Environmental Building News.* The nation's leading source of objective information on green building, including alternative energy and back-up heating systems. Archives containing all issues published from 1992 to 2001 are available on a CD-Rom from BuildingGreen, Inc., 122 Birge Street, Suite 30, Brattleboro,

VT 05301. Tel: (803) 257- 7300. Web site: www.BuildingGreen.com

*Environmental Design and Construction.* Publishes numerous articles on green building. 81 Landers Street, San Francisco, CA 94114. Tel: (415) 863-2614. Web site: www.EDCmag.com.

*Natural Home.* Publishes numerous articles on natural building and healthy building products. Contact them at: 201 Fourth St., Loveland, CO 80537. Web site: www.naturalhomemagazine.com.

## ORGANIZATIONS

**Air Conditioning and Refrigeration Institute (ARI).** Offers information on in-duct air filtration/air cleaning devices. 4301 N. Fairfax Dr., Suite 425, Arlington, VA 22203. Tel: (703) 524-8800. Web site: www.ari.org.

**American Academy of Environmental Medicine.** Contact them for the name of a physician who is qualified to diagnose and treat multiple chemical sensitivity. 10 E. Randolph Street, New Hope, PA 18938. Tel: (215) 862-4544.

**American Academy of Otolaryngologic Allergists.** Another source for names of physicians qualified to diagnose and treat multiple chemical sensitivity. 8455 Colesville Road, #745, Silver Springs, MD 20901. Tel: (301) 588-1800.

**American Society of Heating, Refrigerating, and Air Condition-**

ing Engineers (ASHRAE). Provides information on air filters. 1791 Tullie Circle, NE, Atlanta GA 30329. Web site: www.Ashrae.org.

**Association of Home Appliance Manufacturers (AHAM).** For information on standards for portable air cleaners. 20 North Wacker Drivee, Chicago, IL 60606. Tel: (312) 984-5800, ext. 308. Web site: www.aham.org.

**BuildingGreen, Inc.** Publishes *Environmental Building News*, *GreenSpec Directory* (a comprehensive listing of green building materials), *Green Building Advisor* (a CD-Rom that provides advice on incorporating incorporating green building materials and techniques in residential and commercial applications), and Premium Online Resources (a Web site containing an electronic version of its newsletter, the GreenSpec products database, and more). 122 Birge St., Suite 30, Brattleboro, VT 05301. Tel (800) 861-0954. Web site: www.BuildingGreen.com.

**Conservation and Renewable Energy Inquiry and Referral Service.** U.S. Department of Energy office for information and a referral on air-to-air heat exchangers. P.O. Box 3048, Merrifield, VA 22116. Tel: (800) 523-2929.

**Gas Appliance Manufacturers Association, Inc.** For information on gas heating appliances. 1901 N. Moore Street, Suite 1100, Arlington, VA 22209.

**The Healthy House Institute.** Offers books and videos on healthy building. Contact them at 430 N. Sewell Road, Bloomington, IN 47408. Tel: (812) 332-5073. Web site: http://hhinst.com/index.html.

**Indoor Air Quality Information Clearinghouse.** Distributes EPA publications, answers questions, and makes referrals to other nonprofit and government organizations. Contact them at: P.O. Box 37133, Washington, DC 20013-7133. Tel: (800) 438-4318.

**Multiple Chemical Sensitivity Referral and Resources.** Professional outreach, patient support, and public advocacy devoted to the diagnosis, treatment, accommodation, and prevention of multiple chemical sensitivity disorders. 508 Westgate Road, Baltimore, MD 21229. Tel: (410) 362-6400. Web site: www.mcsrr.org.

**National Association of Home Builders Research Center.** A leader in green building, including energy efficiency. Sponsors important conferences, research, and publications. 1201 15th St. NW, Washington, DC 20005. Tel: (800) 898-2842. Web site: www.nahbrc.org. For a listing of their books contact www.builderbooks.com

**National Radon Hotline.** Calling this number or contacting their Web site will give you access to local contacts who can answer radon questions. Tel: (800) /SOS-RADON. Web site: www.epa.gov/iaq/contacts.html.

**U.S. Consumer Product Safety Commission.** Contact them for information on potentially hazardous products or to report one yourself. CPSC, Washington, DC 20207-0001. Tel: (800) 638-CPSC. Web site: www.cpsc.gov/.

**Wood Heater Program**, U.S. Environmental Protection Agency, For information on woodstoves. OECA/OC/METD, 401 M Street, SW, Washington, DC 20460. Tel: (202) 564-2300.

**Wood Heating Alliance.** For answers to questions about the safety of woodburning stoves. 1101 Connecticut Ave, NW, Suite 700, Washington, DC 20036.

## SUPPLIERS: GREEN AND HEALTHY BUILDING MATERIALS

Because there are many manufacturers of healthy, green building materials, please refer to *Green-Spec, Green Building Resource Guide, Green Building Materials,* or *Sustainable Building Sourcebook*. Below is a list of retailers who sell healthy, environmentally friendly building materials, paints, stains, and finishes.

**Building for Health Materials Center.** Offers a complete line of healthy, environmentally safe building materials and home appliances including straw bale construction products; natural plastering products; flooring; natural paints, oils, stains, and finishes; sealants; and construction materials. Offers special pricing for owner-builders and contractors. P.O. Box 113, Carbondale, CO 81623. Tel: (970) 963-0437. Web site: www.buildingforhealth.com.

**EcoBuild.** This new company in Boulder, Co. works specifically with builders, providing consultation and green building materials at competitive prices. Call David Adamson at: (303) 544-6255. Web site: www.eco-build.com.

**Eco-Products, Inc.** Offers a variety of green building products including plastic lumber. 1780 55th Street, Boulder, CO 80301. Tel: (303) 449-1876.

**Eco-Wise.** Retail store that carries Livos and Auro nontoxic natural finishes and adhesives. 110 W. Elizabeth, Austin, TX 78704. Tel: (512) 326-4474. Web site: www.ecowise.com.

**Environmental Building Supplies.** Green building materials outlet for the Pacific Northwest. 1331 NW Kearney Street, Portland, OR 97209. Tel: (503) 222-3881. Web site: www.ecohaus.com.

**Environmental Construction Outfitters.** Sells an assortment of green building materials. 44 Crosby Street, New York, NY 10012. Tel: (800) 238-5008. Web site: www.environproducts.com.

**Environmental Home Center.** Offers a variety of green building materials. 1724 4th Ave. South,

Seattle, WA 98134.
Tel: (800) 281-9785. Web site:
www.environmentalhomecenter.
com.

**Planetary Solutions.** Long-time
green building material supplier.
Offers paints, flooring, tile, and
much more. 2030 17th Street,
Boulder, CO 80302.
Tel: (303) 442-6228.
Web site: www.planet earth.com.

# INDEX

Grahl, Christine, 137
graywater, 245
green building materials, 243–44
*Green Building Resource Guide,* 69
*Green Spec,* 64, 69
ground-source heat pumps, 141–44
Guelberth, Cedar Rose, 197, 198
*Guidelines for Home Building,* 30, 35

## H

HardSeal, 198
*Healthy by Design,* 195
*Healthy House, The,* 204
*Healthy House Answer Book, The,* 195, 196
heat exchangers, air-to-air, 202, 204
heat gain, analysis of, 231–32, 234
heat gain, reduction of
    external gain, 165–74, 181–82
    internal gain, 164–65, 181
heating degree days, 92–94, 216, 225
heating load
    analysis of, 216, 218, 224
    defined, 90
    passive solar heating and, 89, 91
    sun-tempered design and, 91
heating season, 90
heating systems
    back-up (*See* back-up heating systems)
    passive (*See* passive solar heating)
heat loss
    analysis of, 215–16, 219, 224, 234
    from attic and whole-house fans, 178
    from clerestory windows, 98
    heat-recovery ventilators (HRVs) and, 202
    from skylights, 99
    from thermal storage walls, 118
    from windows, 24, 77–78, 110
heat pumps, 131, 141–45, 181
heat-recovery ventilators (HRVs), 202, 204
Heliodyne PV-powered solar water heater, 148
Hensley, Jay, 157, 162

HEPA air filters, 210
Hermannsson, James, 69
Home Energy Efficient Design (HEED), 233
hot-water heating systems
    baseboard, 131, 133–34
    interior air pollution from, 192, 196
    solar, 7, 131, 145–48
household cleaning agents and disinfectants, 192–93
house plants, 110, 122, 125, 185
housewraps, 73
*Humanure Handbook, The,* 246

## I

Icynene, 67
Ihorne, Jennifer, 181
*Independent Builder, The,* 243
indirect-gain passive solar heating. *See* thermal storage walls
indoor air pollution, 107, 191–211
    air filters to control, 204–11
    eliminating sources of, 195–98
    health effects of, 193–95
    isolating sources of, 198–200
    overview, 191–93
    ventilation to control, 36, 200–204
induced-draft gas furnaces, 138
inert gases, in windows, 76
insulating concrete forms (ICFs), 57–58
insulation, 31–34, 48
    attached sunspaces, 119–22, 124, 125
    basements, 32–33, 52, 54
    ceilings, 60–71
    conventional types, 60–69
    direct-gain passive solar heating systems, 101, 106–107
    energy performance analysis, 232
    envelope, 27, 87, 214
    floor, 60–71, 106, 134 (*See also* subslab insulation)
    foundations (*See* foundations)
    moisture, protection from, 34–35, 71–73, 214
    natural, 69–71
    passive cooling, 173, 184, 186

perimeter, 29, 124, 134
sealing building with, 35
skylights, covers for, 99
subslab (*See* subslab insulation)
sun-tempered designs, 91
thermal storage walls, 115–17
walls, 60–71
windows, 33, 106, 110
wing, 55–56
integrated design, 87–92, 125–27, 213–36. *See also* region-specific design
    direct-gain (*See* direct-gain passive solar heating)
    indirect-gain (*See* thermal storage walls)
    isolated-gain (*See* sunspaces, attached)
    overview, 9–10
    passive cooling, 181–83
    principles of, 10–44, 89–90, 213–14
    sun-tempered designs, 90–91, 218
    team approach to, 10
interior finishes, 105, 115, 196–97
interior partition walls, 30
International Energy Conservation Code, 31, 91, 214
ion generators, 207
irritants, 193. *See also* indoor air pollution
isolated-gain solar designs. *See* sunspaces, attached

## J

Jenkins, Joseph, 246
Judkoff, Ron, 30, 42, 101, 116, 217

## K

Kachadorian, James, 109, 214
King, Bruce, 242

## L

landscaping. *See also* trees; vegetation
    for environment and energy, 247
    for shade, 168, 187, 213
    for wind protection, 40–43, 214

*Passive Solar House, The,* 109, 214–16

*Passive Solar House Basics,* 99

Pearson, David, 240

pellet stoves, 131, 156

pesticides, 193, 197

Pettit, Betsy, 49, 184

photovoltaics (PV), 7, 237

planters, 30

polysiocyanurate foam insulation, 66

Polysteel forms, 57

polystyrene (EPS) foam insulation, 56, 68

polyurethane insulation, 66–69

porches, 187

privacy, 110

*Profit from Building Green,* 181

**R**

radiant barriers, 32, 184, 186

radiant-floor heat, 129, 131, 134–38

radiant wood stoves, 152

radon gas, 194, 199–200

rammed earth construction, 30, 31, 111, 115, 179, 183, 186, 239

ranges, gas, 192, 196

Rastra block, 51

Real Goods, 147–48

rectangular floor plan, 22–23, 36–37, 213

region-specific design, 92–95

    attached sunspaces, 123–24

    direct-gain systems, 104–108

    energy performance analysis, 218–27, 233

    passive cooling, 184–89

    thermal storage walls, 116–17

RESFEN, 84

resource guide, 255–65

rigid-foam insulation, 64–66

    foundations, 52–56

    thermal storage walls, 114, 115

roller shades, 26, 169–70

roof materials for passive cooling, 171–72

roof vents. *See* vents, ridge and soffit

Rosenbaum, Marc, 137

Rousseau, David, 195

Roxul Drain Board, 53

Roy, Rob, 43

R-value, 81, 116

    building envelope, 214–15

    insulating concrete forms (ICFs), 57–58

    insulation, 31, 53, 56, 173

    window shades, 171

    wood, 103

**S**

Safecoat paints and stains, 197

SafeSeal, 198

Schaeffer, John, 147

Schiler, Marc, 43, 168, 185

seasonal availability of sunlight, 11–12, 15, 92–95, 249–54. *See also* weather data files

selective surface foils and paints, 115

shade, provision of, 23–26, 213. *See also* overhangs

    biological shading, 168, 187, 213

    direct-gain passive solar heating, 100, 101, 110

    passive cooling, 101, 166–71, 184–89

shades, 23, 25, 110, 114

    attached sunspaces, 122, 124

    automatic *vs.* manual, 116, 171

    external, 26, 114, 169–70, 184

    interior, 114, 116, 170–71

    mechanical shading devices, 168–71

    skylights, 99

    thermal storage walls, 114

sheep's wool insulation, 69–70

Sheinkopf, Kenneth, 181

shutters

    exterior, 26, 33–34, 169

    internal rigid, 114

    thermo-shutters, 33, 34

Siegenthaler, John, 134

Sikor, Jeannie Legget, 181

site selection, factors in. *See also* orientation of building

    assessing suitability of site, 18

    solar exposure, 10–19

    sustainable buildings, 240

Sklar, Scott, 181

skylights, 9, 23, 88, 98–100, 166–67, 186–87, 222

slab-on-grade foundations. *See* foundations

slope, south-facing, 18–19

smaller homes, 37, 241

solar collectors, 145–46. *See also* photovoltaics (PV)

solar electric systems, 7, 237

solar energy, available, 89, 92–95

solar exposure, site with good, 10–19

solar gain, 214

    defined, 90

    delayed, 111

    direct-gain (*See* direct-gain passive solar heating)

    energy performance analysis, 215–16, 225

    indirect-gain (*See* thermal storage walls)

    isolated-gain (*See* sunspaces, attached)

    orientation of house for (*See* orientation of building)

    overhangs and shading to regulate, 24–26

solar glazing, 9, 23, 77–78, 88–91, 213, 220, 222

    clerestory windows, 96–98

    defined, 90

    direct-gain systems, 95–100, 105

    external heat gain, regulating, 166–67

    glass-to-mass ratio, 30, 102–104, 127

    orientation of house and, 21–22

    skylights, 98–100

    vertical *vs.* tilted, 95–96

solar heat gain coefficient (SHGC), window, 81–82

solar heating, passive. *See* passive solar heating

solar hot-water systems, 7, 131, 145–48

*Solar House, The,* 32

Solar Pathfinder, 18

solar radiation by region, map of, 11–12, 92–95, 227

solar screens, 169–70

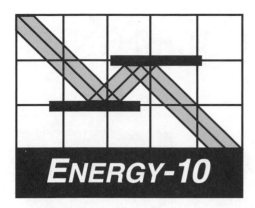

ENERGY-10

## Designing Low-Energy Buildings with ENERGY-10

## What is it?

*ENERGY-10* software is a powerful design tool that analyzes and illustrates the energy and cost savings that can be achieved through more than a dozen sustainable design strategies. Hourly energy simulations help you quantify, assess, and clearly depict the benefits of daylighting, passive solar heating, natural ventilation, well-insulated envelopes, better windows, lighting systems, mechanical equipment, and more.

## Why use it?

This powerful tool helps you make informed decisions about energy performance during the crucial early phases of design, when sustainable building strategies and materials can be integrated at lowest cost. The *Designing Low-Energy Buildings with ENERGY-10* package helps take the guesswork out of integrated, whole-building design.

## How do I get it?

*ENERGY-10* has been developed by the National Renewable Energy Laboratory, with the support of the U.S. Department of Energy. It can be purchased through the Sustainable Buildings Industry Council for $250 (professional), or $50 (student/academic). Site licenses are also available.

## To order on-line visit www.sbicouncil.org.

**Use the special Chelsea Green reference code "Chelsea01"
and receive a 50% ($125) discount on your purchase.**

**For more information** about *ENERGY-10* and other SBIC resources, contact:

Douglas K. Schroeder, *Associate Director*
Sustainable Buildings Industry Council
1331 H Street, NW, Suite 1000
Washington, DC 20005
(202) 628-7400 ext. 210
Dschroeder@sbicouncil.org

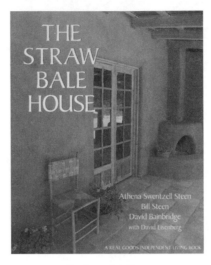